THE ROOTS OF VOCATIONAL EDUCATION

EDITED BY
J. C. WRIGHT

THE ROOTS OF VOCATIONAL EDUCATION.
By WILLIAM P. SEARS, JR., PH.D., Instructor in Mathematics, Marquand School for Boys, and Central Branch Y. M. C. A. Evening School, Brooklyn, N. Y. 310 pages. Cloth.

FOUNDATIONS OF INDUSTRIAL EDUCATION.
By F. THEODORE STRUCK, PH.D., Head of Department of Industrial Education, Pennsylvania State College. 492 pages. 5½ by 8. Cloth.

METHODS AND TEACHING PROBLEMS IN INDUSTRIAL EDUCATION.
By F. THEODORE STRUCK, PH.D., Head of Department of Industrial Education, Pennsylvania State College. 214 pages. 5½ by 8. Cloth.

GUIDANCE AND EDUCATION OF PROSPECTIVE JUNIOR WAGE EARNERS.
By FREDERICK M. TRUMBULL, Vocational Director, Rockford, Ill. 298 pages. 5½ by 8. Many recording forms. Cloth.

THE ADMINISTRATION OF VOCATIONAL EDUCATION.
By J. C. WRIGHT, Director, Federal Board for Vocational Education, and CHARLES R. ALLEN, Editor and Educational Consultant, Federal Board for Vocational Education. 436 pages. 5½ by 8. 16 diagrams, 5 charts. Cloth.

THE SUPERVISION OF VOCATIONAL EDUCATION.
By J. C. WRIGHT and CHARLES R. ALLEN. 415 pages. 5½ by 8. 34 diagrams. Cloth.

EFFICIENCY IN EDUCATION.
By J. C. WRIGHT and CHARLES R. ALLEN. 405 pages. 5½ by 8. 6 diagrams. Cloth.

EFFICIENCY IN VOCATIONAL EDUCATION.
By J. C. WRIGHT and CHARLES R. ALLEN. 443 pages. 5½ by 8. Cloth.

PUBLISHED BY JOHN WILEY & SONS, INC.

THE ROOTS OF VOCATIONAL EDUCATION

A Survey of the Origins of Trade and Industrial Education
Found in Industry, Education, Legislation
and Social Progress

WILLIAM P. SEARS, Jr., Ph.D.
INSTRUCTOR IN MATHEMATICS, MARQUAND SCHOOL FOR BOYS
AND CENTRAL BRANCH Y. M. C. A. EVENING SCHOOL,
BROOKLYN, NEW YORK

NEW YORK
JOHN WILEY & SONS, Inc.
London: CHAPMAN & HALL, Limited
1931

COPYRIGHT, 1931,
BY
WILLIAM P. SEARS

ALL RIGHTS RESERVED

Printed in U. S. A.

THE HADDON CRAFTSMEN
CAMDEN, N. J.

AUTHOR'S PREFACE

America today is one of the foremost of industrial nations. Among the other nations of the world few can compete with it in the quantity and quality of industrial products. Even after the domestic wants have been satisfied, America can produce to such an extent that foreign markets can be supplied, if not dominated.

In reaching this pinnacle of industrial supremacy, America has undergone tremendous changes since its discovery. From a vast trackless wilderness in the sixteenth century it has reached a position in the twentieth century comparable, if not superior, to the positions held by the more advanced of the ancient nations of Europe.

The great changes which have occurred have been felt in all spheres of America's activities. Industry, politics, education, religion, philosophy, and social life have evolved in this short space in an amazing manner. Even today change is ever taking place.

A cross section of the schools of modern America presents a vastly different picture from a similar survey at the close of the seventeenth century. The development of the country's natural resources, the influences of the Industrial Revolution, the opening of the great west and the evolution of industry have left their mark upon education as well as upon the other departments of social life.

The changes in education have been deeply rooted in the changes in industrial life and scientific thought. The change has been more fundamental than a mere change in pupil enrolment. There has come a realization of a broader and deeper understanding of the purpose and function of education. There has come a change in values. The idea that there are different types of education for different types of individuals has gained ground. The specialization which has resulted from the adoption of the factory system of production has been one cause for this breaking up of the courses of instruction. Another cause has been the realization of the existence of individual differences. Applied psychology, through its many and varied research undertakings, has established tests and measurements, which many believe can be of assistance in guiding an individual along the questions of choice and progress in his life work. Education has come to mean the harmonious development of the whole person. The fact that the person must be a wage earner, a fact often overlooked by academic systems of education, is a keystone in the structure of modern education. The trend in American life has been toward industrialism, consequently the trend in education has been toward training for industrial pursuits.

It is the purpose of this study to seek the origins and beginnings of vocational education. An attempt has been made to discover the roots of organized and controlled trade and industrial education.

Vocational education is a comprehensive term and includes professional, commercial, agricultural, domestic and industrial training. The emphasis in this study will be placed upon the industrial phase of vocational education. Industrial education means the complete and appropriate education of industrial workers of whatever grade. It means the preparation of the youth for positions in the industrial world; it means the further development of industrial workers for positions of higher rank and greater service. It means the enrichment of the individual by the development of skills, techniques, and job intelligence. It means the preparation of workers for positions of leadership. It means, also, the training of workers to help them overcome, through short intensive courses of a "pusher" type, those difficulties arising from changing industrial processes or from some lack in their previous training.

It has been found necessary to present a survey of vocational education through the ages, a résumé of the work of educational pioneers, and a brief summary of the industrial history of the

United States. Also included is a digest of the work of the various social agencies which have been instrumental in assisting the movement for vocational education.

And now the *raison d'être* for this treatise. There have been published countless industrial histories, many histories of education, scores of studies of the labor problem, and volumes recording the legislative acts of the various governing bodies. There has been no volume that has attempted to include the significant facts culled from the spheres of industry, labor, and legislation. It is the purpose, then, of this work to gather together the influences which have in the past reacted upon vocational education. From a study of economic, social, and educational conditions an attempt has been made to correlate changes and developments in vocational education.

It is the hope that this study may be of assistance to those interested in education, and more especially industrial and trade education. Year after year hundreds of teachers of trade and industrial subjects flock to the various extension divisions of universities and schools of education. Year after year the normal schools and teacher-training centers enroll men and women who have chosen the teaching of boys and girls for positions in industry as their life work. Those teachers expect to be leaders in the field of vocational education. They seek to improve themselves professionally by gaining a universal appreciation of the field of service to which they are devoted. They must understand the sociological basis upon which vocational education rests. They must be aware of the great story of the evolution of industry. They must know the influence of the outer world upon vocational education. In the past they have wasted much effort and time in seeking their material from various sources. It is hoped that in this study they can find enough material to guide them and point the way to more specialized and detailed paths of research.

For those who might select this as a text for some such course, it is suggested that this book be used as the regular reading material and that the individual student pursue some one definite root and seek out, in as specific a way as the facilities at his disposal allow, noting its contributions and relationships.

At the beginning of each section of this work credit is given for those whose works have been drawn upon. Of Charles A. Bennett and Edward S. Cowdrick especial mention may be made. Rather than delay the progress of the work with countless footnotes a blanket acknowledgment to them is preferable.

Finally a debt of gratitude must be acknowledged to Dr. Ralph E. Pickett of New York University whose devoted interest and indefatigable labor has made this work possible. As friend and teacher he has been a constant source of inspiration.

WILLIAM P. SEARS, JR.

NEW YORK CITY
September, 1930.

EDITOR'S PREFACE

All things must have a beginning and if they endure for any great length of time there must be a good reason for their continued existence. The author of "The Roots of Vocational Education" has sought to unearth the principal beginnings in this field of education, and to explain why it is of such importance in our present civilization.

Most of us are accustomed to think in terms of twenty or thirty years, when we speak of the history of vocational education. As a matter of fact, its history antedates all other educational beginnings.

The author is to be commended for the zeal with which he has pursued the many ramifications of the subject, and the interesting manner in which they are presented to the reader.

J. C. WRIGHT.

CONTENTS

PART ONE
HISTORICAL BACKGROUNDS 1

PART TWO
THE EDUCATIONAL EVOLUTION 63

PART THREE
THE CONTRIBUTIONS OF SOCIAL AGENCIES AND EDUCATORS TO VOCATIONAL EDUCATION 159

PART FOUR
ORGANIZED BUSINESS AND VOCATIONAL EDUCATION . . . 211

PART FIVE
VOCATIONAL EDUCATION IN CANADA 227

PART SIX
VOCATIONAL EDUCATION AND THE FUTURE 241

PART SEVEN
MILESTONES IN THE DEVELOPMENT OF VOCATIONAL EDUCATION 249

PART EIGHT
THE ROOTS OF VOCATIONAL EDUCATION—A SUMMARY . . 275

PART NINE
CLASSIFIED BIBLIOGRAPHY 293

INDEX 305

Part One

HISTORICAL BACKGROUNDS

Part One of this study deals with the historical backgrounds against which the forces that have created vocational education have played. It consists of a résumé of the industrial and social history of England and the United States.

Many industrial and political histories have been consulted in the compilation of this material. E. L. Bogart's "The Economic History of the United States" has furnished numerous items, but to Edward Cowdrick's volume, "The Industrial History of the United States," the greatest debt is due. This abundant source has been constantly drawn on during the construction of this treatise. In many instances actual footnote references have been omitted. Statistical data, both numerical and descriptive, have been taken from Cowdrick, who mentions the various United States Census Reports and the Statistical Abstract as original sources.

THE EVOLUTION OF INDUSTRY

English Backgrounds

Introduction.—Thorndike, in the opening sentence of his "Educational Psychology," says: "The arts and sciences serve human welfare by helping man to change the world, including man himself, for the better." These changes in man's material environment in the past century have been striking. Even a cursory glance at the economic and industrial structures of present-day society will reveal a picture radically different from the pictures of former times.

The evolution of industry from earliest times to the present is the story of man's increasing power over nature and his greater capacity to control and direct the forces of nature for his own use.

English Industrial History a Background.—The explanation of American industrial growth is made easier by a study of its origins in English industrial and economic history.

England in 1066.—England was an agricultural country when William the Conqueror led the Normans thither in 1066. The Anglo-Saxons cultivated the soil, grew and harvested the crops, and tended to the distribution of the earth's products. These people were held to their land by the capital invested therein, i. e. the crops, tools and the land itself. Trade, in the form of the exchange of agricultural surplus, was a characteristic of this period.

The Manor System of Old England.—Manorial England from 1066 to the sixteenth century remained in this stage of economic development. Rural England was composed of a number of self-supporting units called manors, loosely held together by a weak central government. These manors ranged in size from a hundred acres upward.

Each manor was a world unto itself and may be considered in two aspects:—(1) the manor, and (2) the vill. The manor consisted of the manor house, the lord's demesne, and the manor court, while the vill consisted of the village and the open fields.

The manor house was the center of the manor, and in it ruled the lord of the manor. The lord might be the king, or a noble, abbot, or squire.

The manor was profitable to the lord in various ways. He received rents and payments from his tenants. He also received profits from the farming of the demesne farm. In this case the lord was simply a large farmer with a supply of free labor, for he received payment from his lower tenants in the form of service. Also, he received revenue from the manor court.

The demesne was a farm which the lord kept for himself, cultivated by the villeins and cotters.

At the manor the tenants gathered from time to time for judicial and regulative action. Questions of transfer, fines, inheritance, and election of petty officers were settled in the "court baron", while punishments, breach of contracts, and breaches of assizes were settled in the "court leet". The courts were presided over by a steward, acting for the lord, and all adult males had to attend.

The Division of Labor on the Manor.—A further division of labor was manifested by the presence of several petty officials. The reeve was an officer of the manor chosen by the lord, or elected by the tenants, whose duty it was to run the manor in the absence of the lord. He was required to know the qualities of soil, to keep the accounts of the manor, to manage the household servants of the lord, and to attend to various menial tasks. Every manor also had a miller, who paid a high rental to the lord. At his mill, all the grain produced on the manor had to be ground. The parish priest looked after the spiritual well-being of the people of the manor. The lord's implements of farming and warfare were cared for by a blacksmith, who also made and repaired tools for the tenants. These miscellaneous officers were universal fixtures on the manors of medieval England.

THE MANOR SYSTEM

Farming on the Manor.—The vill consisted of a village and the surrounding open fields of farm land. The lands of the cultivators lay together in large tracts, and were cultivated in strips of acres, half acres, or quarter acres. The usual method of farming was the wasteful three-field system. The plow land was divided into three great fields. These were inclosed by fences. One field was sown to wheat in the fall; one to rye or barley in the spring; and the third lay fallow to recuperate. The next year this field would be wheat land, while the old wheat field would raise the barley, and so on.

The tenants who made up the vill ranged from holders of as much as one hundred acres down to cotters who held but a house and perhaps a half acre. All these tenants held their land from the lord of the manor and made payments to him in produce or service, or both. The free tenants, in the latter part of the period, paid the lord in money; the villeins paid him in boon work and week work. Special assessments and taxes were also collected at various occasions, as, for example, the marriage of the lord's daughter, the knighting of his eldest son, or the inheritance of a holding by the son of a deceased tenant.

The free tenants were those freemen who held the largest amounts of land and who paid in money rather than in service.

The Serfs.—The villeins, or serfs, were bound to the soil and could not leave the manor. This class was largest, in point of numbers. They paid rent in the form of service on the demesne farm, and by hard work, both for the lord and for themselves and family, managed to keep body and soul together. The cotters finally became one with this class.

However, besides the possession of their strip of land and the right to pasture on the common greens, there were some slight alleviations of the burdens of the cotters and the villeins. The lord provided food for one or two meals a day at the time of the boon and week work period. Also, once a year the lord gave a specific amount of grain, fish, and beer.

The Manor System a Root of Class Distinction.—From a social point of view this epoch, characterized by the agricultural economy of the manor, made for a class distinction on the manor, and for the complete separation of the people of any one manor from the outside world. This proved to be an obstacle to progress and national feeling. It was an arch-enemy of commerce and trade, as well as of industrial and educational development.

The Decay of the Manor System.—It was not until the fifteenth and sixteenth centuries that England came out of this period of agricultural development and economy. The leading causes of the decay of the manor system may be summarized briefly. One of the chief causes was the substitution of money payments for service payment. The increased amount of money also allowed the lord to hire laborers who were more efficient than the serfs. The Black Death of 1348 cut the population of England in half. This decreased the number of workers, thus increasing the demand for labor. The increased amount of money per capita led to increased prices. This also increased the commutation of services.

The next step was the passage of the Statute of Laborers which resulted in keeping the price of labor down, in attempting to maintain reasonable prices for food and in providing that no alms be given. The Hundred Years War was causing a drain upon the money of the land, and taxes were being levied. All these causes resulted in the Peasants' Revolt of 1381. And, although the results of this revolt were nil, as far as immediate changes were concerned, it showed the growing unrest in the labor class.

Another cause of the breakdown of the manorial system was the realization on the part of the lords that sheep grazing could be made more profitable than farming. The rise in the price of labor and the increased demand for wool gave this movement, called the "enclosures", its impetus. The enclosures were the first result of the development of the wool industries. Less labor was required for sheep raising, but a greater amount of land, than for farming. The conversion of farm land into grazing land brings to a close this period. Thousands of small farmers found themselves without farms and without work upon the farms of larger holders, who now turned their attention to sheep raising. These dispossessed farmers wandered about for a short period, until they finally found themselves in the growing cities. This story properly belongs to the story of the Industrial Revolution. Richard Ely, in his "Outlines of Economics", p. 38, lists the forces of change at work at the end of the manorial period thus:

"(a) A rapid growth in the number of free tenants; (b) the commutation of customary services into fixed payments of money or kind; (c) The appearance of a class of agricultural laborers dependent on the wages they received."

The Handicraft Era in England. The Guilds.—The handicraft stage of economic development fills that period beginning with the rise of towns as centers of handicraft and trade until the introduction of power-driven machinery. This includes approximately the period from 1300 to 1800. After the manorial system had developed, there grew up the guild system in the towns. Merchant guilds received trade monopolies from the king, in exchange for allegiance and services. These guilds regulated the buying and selling of goods, the times and places of sales, prices, and fair dealing. The rapid growth of the towns led to the increased number of handicrafts, the artisans of which organized the craft guilds. The craft guilds promoted honest work, fraternal and social activity, and standards of production. They regulated weights and measures, forbade the adulteration of goods, and, in many cases, prohibited night work. One of the most important functions of the craft guilds was the control of admission to the craft and the setting of the period and method of apprenticeship.

When the manorial and guild systems decayed, there grew up the so-called domestic system. This stage of economic evolution persists on a limited scale today in England and other European countries. It flourished as the characteristic form of English industry in the sixteenth and seventeenth centuries. Under this system each household was a production center and all members were producers. A further class distinction evolved when the function of merchant and worker separated, and the middle class furnished the raw material for the households.

Labor in the Fourteenth Century.—Labor benefited greatly from the later effects of the Black Death, which visited England in 1348. The working class, living in congested centers in filth and poverty, was, it is estimated, cut in half. The free artisans and serfs who had lost their masters took advantage of the enhanced price for service and the favorable labor demand. The following sequence is similar to that which took place on the manors. The Statute of Laborers, of 1351, ordered workmen to accept work when it was offered and at definitely established wages. Punishment, either of the stocks or by imprisonment, for failure to comply with these laws was enforced.

The workers in the towns and the serfs on the manors were bent under these oppressive acts and the natural result was a combination for redress of grievances. The Peasants' Revolt in 1381 was the natural protest against unfair laws, oppressive conditions, and the high prices of commodities, with which the English working class had to contend.

The improved condition of labor in the century following has earned for this period the title of the "golden age" of the English laborer. The developing wool industry, together with the high prices paid to wool workers, separated the people from the land. Next followed a period of regulation of industry under the Tudor rulers. The regulations were national in scope and extended into the fields of weights and measures, coinage, conditions of labor, etc. A reenactment of the Statute of Laborers called the Statute of Apprentices was passed in 1563, to remain on the books until 1813. E. P. Cheyney, in his "Industrial and Social History of England", p. 156, says of this act:

"It made labor compulsory, and imposed on the justices of the peace the duty of meeting in each locality once a year to establish wages for each kind of industry. It required a seven years' apprenticeship for every person who should engage in any trade, establishing a working day of twelve hours in summer and during daylight in winter; and enacted that all engagements, except those for piecework, should be for the year, with six months' notice at the close of contract by either employer or employee. By this statute all relations between

THE GREAT INVENTIONS

master and journeyman and the rules of apprenticeship were regulated by the government instead of by the individual craft guilds."

Mercantilism, and Economic Philosophy.—The economic theory of the time was expressed in the Mercantile System, which stood for the encouragement of an excess of exports and the accumulation of precious metals. Other interesting developments of the time were the granting of monopolies and the encouragement of the immigration of foreign artisans to introduce and establish new industries.

The Industrial Revolution.—The fifth stage in economic evolution is the modern industrial system. The Industrial Revolution changed the entire industrial make-up of England. This change may be termed a change from mercantilism. It was accompanied by a rapid growth of population, together with a relative and absolute decline of the agricultural population. It was a change from the domestic to the factory system.

Three Sources Contain the Causes for the Industrial Revolution.—The causes of the Industrial Revolution can be classified under three types: First, the objective causes, which include the development and invention of new types of machinery and the associated development in the uses of capital; second, the subjective causes, residing in changes in philosophy and mental attitude; and, third, the economic causes, especially the development and extension of English trade and the growth of demand centers overseas in the form of dominions, such as America.

The Great Inventions of the Eighteenth and Nineteenth Centuries.—The root of the Industrial Revolution lies in the series of great mechanical inventions occurring between 1760 and 1820. The industrial, social, and moral effects of these inventions are almost incalculable even today. Their history is a fascinating study in itself. The increased production demands upon the antiquated methods of manufacture and transportation could not be met. The England of the sixteenth century could utilize the domestic system for the satisfaction of its wants; the England after the eighteenth century could not. The domestic system proved insufficient and inadequate to meet the needs of the commercial and industrial era upon which the nation was about to enter. Invention led England into a new and different industrial era. In 1738, Kay invented a flying shuttle which facilitated the weaving process and multiplied the demand for yarn. In 1764, Hargreaves' "spinning jenny" improved hand spinning. There followed Arkwright's water frame in 1765. Next, in 1779, came Crompton's spinning mule, which combined the inventions of Arkwright and Hargreaves. This led to a demand for more thread. Cartwright's loom of 1784 began the modernization of weaving. The invention of the cotton gin by Whitney in 1793 opened up the American cotton fields.

Next came developments with power. Watt improved the steam engine, and it was applied to the mining and manufacturing industries. The use of the improved steam engine led to its adoption in the textile field in 1785. The development of machine tools came as a next step in answer to the demand for machinery, which was too great to be satisfied by hand production.

Other interesting lines of development at this time were the improvement of roads, the building of canals, and the development of steam locomotion. George Stephenson in 1814 applied steam power to land transportation, and in 1825 the first English railroad was opened for traffic. These great inventions placed England at the head of the industrial world.

The Results of the Industrial Revolution.—The results of the Industrial Revolution were the factory system, increased production, a sharp distinction between capital and labor, widened markets, the increased division of labor as manifested in modern specialization, the revival of the enclosures, the decay of the domestic system, the origin of labor unionism, and the *laissez-faire* policy in industrial and social control.

The small handicrafters were put out of business by the factories. These small producers flocked to the cities, where they found that machinery had taken the place of manpower in many lines of trade. The result was an oversupply of labor, and a conflict among the unemployed. The application of mechanical power, the use of larger accumulations of capital, and the collection of scattered laborers into centralized and strictly regulated establishments are the characteristics of the factory system. Factory conditions were unsanitary and inhumane. Women and children were employed for fifteen hours or more a day. The factory buildings were old, dark, and unsanitary. The life of an American slave has been described as pleasant as compared to the lot of the English labor class of the nineteenth century. The helplessness of labor at this time was one of the causes of unionism.

The Development of Labor and Capital as Opposing Forces.—Labor and capital assumed new positions. They became fixed classes. The payment of wages for work done and elimination of personal ties and obligations between the worker and his employer were characteristics of the period. The worker was separated from the ownership and control of the machinery and tools of production. The capitalist owned the tools and machinery. This also distinguishes the period from the handicraft era, when the worker owned the tools.

Watkins, in his "Introduction to the Study of Labor Problems", page 16, states:

"From the standpoint of the wage-earning classes the most significant effect of the Industrial Revolution is this: The major portion of the workers do not possess the opportunity to accumulate the sum of capital necessary to organize and operate modern industrial and business establishments. The majority of the population, therefore, is relegated to the permanent, subordinate status of a wage-earning class. Distinct economic classes of capitalists, entrepreneurs, and wage-earners arise, with interests more or less antagonistic."

Trade Increased After the Advent of the Factory System.—Markets were widened, and competition became more universal. Increased production led to the extension of commodity markets overseas. Steam transportation and improved communication, just developed, linked together national markets and, later, world markets. This increased international competition. It also allowed new countries, such as America, to secure their labor supply from all nations.

Minute Division of Labor a Result of the Modern Machine.—The complexity of the newly created machines led to an increasing specialization of jobs, with the result that few workers performed all the processes in the production of a commodity. In the modern factory system, each worker is usually responsible for only one operation upon a given commodity. The question of the subordination of the worker to the machine, and the narrow specialization of jobs, is important to our study, for it is pertinent in the discussion of trade training for industrial efficiency.

The Eighteenth Century Revival of the Enclosures.—The enclosures were revived in the eighteenth century, due to the high prices that could be received for truck garden products to feed the increasing city populations. Hand labor could not compete with the power-driven machinery of the factories, and the domestic system was broken. The hand workers who could not adapt themselves to the factory regulations and methods had to accept inferior forms of work or sink to pauperism in the rising industrial centers. Many became farm hands, only to lead a life of toil and virtual serfdom on a truck farm for almost no wage at all.

The Employer and the Employee.—The rise of the two classes, the employer and the employee, as distinct groups, hostile, if not antagonistic, was another result of the new economic order. The workers now became mere cogs in the wheels of a capitalistic machine. They had no voice in the administration or direction of the plant.

The Employees Unite.—This breach led to the formation of permanent organizations among the workers. These societies differed from the medieval guilds in that they included wage earners only, not employers as well. The increased production brought together large numbers of workers, who, perceiving their common interests, banded together for mutual aid and protection. Collective bargaining became a prominent weapon of the united laborers. This form of procedure alarmed both employer and Parliament. The result was the passage of the Combination Acts from time to time during the early part of the nineteenth century. These acts had for their core the prohibition of artisan organizations for the changing of the rate and conditions of labor. These Acts declared illegal the attempts of workers to change hours of labor, to improve wages, and to refuse to work. Workers who entered a combination or union could be punished upon conviction by imprisonment for two months. Still another obstacle to unions was the fact that under common law continued attempts to change hours, prices, and methods of apprenticeship were acts in restraint of trade. Popular opinion, too, condemned workers' unions. This state of hostility toward unionism lasted until well after 1825, when the first steps were taken to lift restrictions on unions.

Laissez-Faire, a Political Philosophy.—England, during the Industrial Revolution, changed her entire policy toward the economic and social life of the people. The first evidence of this was in the growing dislike and disregard for the policy of governmental regulation and limitation of activity.

The Industrial Revolution brought with it a philosophy of individualism, and the people seized upon the idea of freedom as a cure for all ills. This policy is summed up in the *laissez-faire* doctrine in economic and social thought. Adam Smith was a leader of the *laissez-faire* attitude in economics. Soon Parliament was converted to the *laissez-faire* policy as one of ease and necessity. In 1811, Parliament stated that no interference of the government with the freedom of trade, or with the perfect liberty of every individual to dispose of his time and labor in the way and on the terms which he may judge most conducive to his own interests, can take place without violating general principles of the first importance to the prosperity and happiness of the community. There followed a rapid shelving of old restrictive laws. From 1813 to 1846 the Statute of Apprentices' clause on wage setting, the East India monopoly, and the tariff restrictions upon foreign trade went by the board. Thus the restrictive regulations of the sixteenth and seventeenth centuries were practically discarded.

This doctrine of individualism constitutes a social philosophy which persists today. Absolute freedom, enlightened self-interest and universal free competition are some of its characteristic phases.

Population Changes in England After the Industrial Revolution.—One social effect of this gigantic industrial change was a rapid growth in population. In 1700 England had a population of about 5,000,000; in 1750, about 6,500,000; in 1800, about 9,000,000; in 1850, about 18,000,000. The population in 1920 was about 45,000,000. The Industrial Revolution, in reaching England first of the great nations, gave to Great Britain a great lead in the race for industrial supremacy.

The Labor Problem Emerges.—The results of the Industrial Revolution, however, worked severe hardships on the small producers and wage-earners, who could not make the readjustments necessary for the transition from the domestic system to the factory system. Women and children were employed extensively in the new factories. The case of operating the new machinery accounts for the persistent employment of this type of labor.

Factory hours were continuous and long. The working day of fourteen hours was not unusual. Employment was not steady, but fluctuated with seasonal and periodic depressions. Factory towns grew up rapidly in the north and west of England. There was no provision for drainage, water, or other sanitary necessities. Rents were high in these congested and filthy centers of population. Transportation was poor, hence the workers had to live within a short distance of the place of employment. Education and amusement for the working classes were of the lowest order. Unrest, unemployment, and illiteracy, together with poor living conditions, were the early manifestations of the machine age.

Governmental Restrictions Blast the *Laissez-Faire* System.—Unrestrained individualism, however, was soon to run its course. The Manchester Board of Health early made a report on the health and living conditions of the city, an evidence of an early attempt at social control. Parliament next passed an act to the effect that children under seven should not be employed over seventy-two hours a week. Next came the example of Robert Owen, who ran a model factory. He treated his employees as human beings. The influence of these events soon reacted upon public opinion, and, together with the extension of the franchise, led to a social consciousness expressed in humanitarian attitudes, ideals, and practices.

Conclusion.—Watkins, in his "Introduction to the Study of Labor Problems," page 22, sums up the evolution of industry as follows:

"In the course of economic evolution the status of the worker has changed from slavery to serfdom and from serfdom to freedom of contract. With the increasing requirements of capital for the organization and operation of business under modern industrialism, the worker has been separated from the ownership and direction of industry, with the result that it is extremely difficult to emerge from the wage-earning class. This position of industrial subordination has led to the recognition of an identity of interests, followed by organization of wage-earners for mutual protection. The emphasis of the modern factory system has been upon the adjustment of the worker to the machine, and this has resulted in serious economic and social problems. Self-interest rather than mutual aid has been the dominating motive in economic life, and to this fact the general conflict of economic interests can be attributed. Industrial selfishness, engendered by the desire for gain, has given rise to numerous labor problems in the Old World."

THE INDUSTRIAL EVOLUTION

The United States

Introduction.—The story of the discovery and settlement of the New World is now fast becoming ancient history. The industrial development of the country, however, has received but little attention as compared with the political development. It is not necessary to recall the reasons for the discovery, or the story of the early settlements.

The Nature of Early Colonial Agriculture.—Agriculture was the first industry carried on by the settlers of America. In New England, the small farm prevailed; in the South social ideals and different agricultural methods established the plantation type of large farm. The crop of the colonists was Indian corn. The culture of this product was taught them by the Indians. Imported seed grains were also experimented with. Some were successful; others did not prove themselves adaptable to American soil and climatic conditions. The middle colonies raised wheat, too, fairly successfully. Silkworms and grapes were unsuccessfully attempted in the South, only to be replaced by rice and indigo and, in Virginia, by tobacco. Cotton was not grown for many years.

Other Colonial Industries.—Live stock raising was another enterprise of the colonists. In the northern colonies, lumbering and shipbuilding became natural developments of the magnificent forests. Fishing, off the New England coasts, proved a lucrative employment, while fur trading with the Indians and trappers brought large returns.

England Discouraged Colonial Manufacture.—Manufacturing throughout the colonial period was of minor importance. England, by restrictive regulations, discouraged colonial manufacturing. The colonies were used as a source of raw materials, while England exported to them practically all of their manufactured articles. Even after 1776 the growth of American manufacturing was slow, and, until about 1820, the country still depended upon European imports of manufactured goods.

Early Colonial Manufacturing.—E. S. Cowdrick, in his "Industrial History of the United States," page 39, states:

"It is likely that the earliest settlers brought spinning wheels and looms from England, and that from the founding of the colonies cloth was woven for clothing. For finer textiles, however, dependence was placed upon England. There is a tradition that a spinning mill was built at Salem in 1640. Certain it is that in 1643 Pastor Ezekiel Rogers and his followers, who had been expelled from England some years earlier, set up a woolen mill at Rowley, Massachusetts. When the English government prohibited the export of wool, Massachusetts encouraged the raising of sheep in order to obtain a home supply. A thriving wool industry was built up when, in 1700, Parliament repealed the export duties on English woolen cloth. This made it more profitable to buy from England than to manufacture cloth, and the infant industry declined. A certain amount of 'homespun,' however, continued to be produced."

In this early colonial industry we may see, in miniature, some of the effects of legislative acts in relation to industry. In all ages business has been sensitive both to encouragement and discouragement in the form of public enactments. A survey of industrial development is futile if it overlooks the importance of those influences from the outside.

Metal-Working in Early America.—In Virginia, as soon as Jamestown was founded, iron ore was discovered. A smelting furnace was erected on Falling Creek, a branch of the James River, by workmen in 1619. This industry had a horrible ending, when after a few years of operation, it was destroyed by the Indians in one of their periodical massacres of the workmen and their families. Cowdrick ("Industrial History of the United States," page 40) mentions that iron was likewise discovered in Massachusetts in 1630, and a furnace erected in 1643 by John Winthrop at Lynn. This plant manufactured about seven tons of iron a week.

In 1648 copper was found near Salem, but small demand for this, in addition to small supply, made the copper industry of even less importance than the iron industry.

Colonial Labor. The Indentured Servant.—To conquer the wilderness of the American continent required a labor force. In the North, the landowners and their families cultivated their small farms with the occasional assistance of a hired hand employed at a daily wage. Industry in the North, carpentry, shipbuilding, ropemaking, and the rest, also took merely the labor of the entrepreneur. Labor, in the South, however, was a different problem. The extensive plantations made additional laborers a necessity. Expanding enterprises demanded a steady and sure labor supply. The manufacturing industries of the middle colonies required more hands than could be furnished by the family of the entrepreneur.

In England, at this time, the poorer classes were faced with pauperism, due to the failure of the majority to adapt themselves to the factory system. These maladjusted individuals sought relief from their woes in the New World. They flocked to America and became indentured servants; that is, they contracted to work for a term of years, without pay, for anyone who would pay their passage across the Atlantic. Also, many criminals and vagrants were sent to the colonies by English law courts. These sources of labor were common and adequate in the northern part of the country, but were insufficient to meet the needs of the southern plantations. The enterprisers in the colonies found to their joy that the board and keep of indentured servants was small as compared to the returns from their labor. The result was a constant demand for more of these workers. In England, to meet this demand, unscrupulous agents resorted to kidnapping and other foul methods in order to supply the colonists.

Law soon defined the period of service and the rights of the indentured servants. Many of these people were skilled artisans, who when freed, became wealthy and respected members of society. Since the greater number of these servants were of the same race and color as their temporary masters, no social stigma was attached to former indentured servants, and class distinctions from this cause did not arise. This redemptionary system was, despite its many abuses, a useful and purposive institution. It answered a demand for cheap and abundant labor at a period when labor was the principal need of the colonists. It also afforded a means of transportation to America for large numbers of desirable people who later became ambitious and energetic citizens.

Negro Slavery in America.—Quite different was the system of negro slavery. The enslaving of the more backward races was a customary practice at this period, and it involved no ethical principle. The introduction of slavery had widespre' social and economic effects on the history of the colonies. the time of the landing of the first negroes, 1619, there v   .o machinery in use. Slave labor proved to be as efficien⁺   .ree labor, with the added attraction of no wage price.

During pre-Revolutionary times, slavery spread   ιgh the southern and middle colonies. In the early ye⸍   ιe Royal African Company of England furnished the slaves, but after 1688 slave trade was thrown open and was engaged in by New England merchants, with great profit. A veritable trade cycle developed. The New Englanders imported molasses from the West Indies. This they manufactured into rum. The rum was shipped to Africa where it was exchanged for negroes. The negroes were then transported to the West Indies where they were sold as slaves. The purchase of molasses then started the cycle once again. In 1808 the importation of slaves from Africa was prohibited by Federal Statute.

Slavery became the prevailing labor system of the South. The application of slave labor to the cotton industry discouraged free white labor and thus left the field open to slave labor. As the cotton growing developed and spread, the demand for slaves grew apace; the result was the illicit slave trade, which is a familiar story in American history. This system spread and continued until 1863, when Lincoln issued the famous Emancipation Proclamation pronouncing that "all persons held as slaves ... are and henceforth shall be free." The status of labor under slavery was that of property.

Population Growth in the Colonies.—Cowdrick, throughout his text and particularly on page 46, calls attention to the oft-forgotten fact that the growth of the colonies was a slow one. From 1607 to 1775 is a longer period than from 1775 to 1930. One of the developments was the growth and spread of population. From a few settlers in 1620, the population grew to about 300,000 in 1700, of which 115,000 were in New England; 100,000 were in the middle colonies, while the rest were in the South.

By 1775 the population exceeded 1,500,000. New England had 475,000; the middle colonies had 407,000; the South, 718,000. Of this population, there were some 275,000 negro slaves. Boston and Philadelphia had each a population of about 25,000; New York was a town of 17,000.

Early Monetary History.—Little coined money had been brought from Europe. Barter was the usual mode of business transaction. Wampum, tobacco, and corn passed as currency at one time or another.

In 1690 Massachusetts found herself unable to pay her soldiers, who had just returned from the siege of Louisburg. The legislature of the colony issued bills of credit to the sum of £40,000. There was no interest or date of maturity upon these bonds, and there was no security except the promise of the colony to pay. The government accepted them in payment of taxes; this kept their value up. In 1711 another issue of £40,000 was followed by similar issues in other colonies. At the end of the colonial period the country was flooded with paper money of various degrees of worthlessness.

The American Revolution.—Then came the Revolution. The causes of this war are listed in all American histories in the fullness required for their treatment. Suffice it here to note a few of the economic causes. The idea of industrial freedom was as prominent as that of political freedom in the minds of many colonial leaders. The aim of the mercantile theory, referred to in the summary of English industrial evolution, was

PARLIAMENT PURPOSED A SYSTEM OF TAXATION 21

to build up for England a strong nationalism and a powerful nation, through shipping and commercial advantages. Regulations were framed so as to amass a domestic wealth which could be taxed and thus provide for national defense and aggression. England enforced these policies in her dealings with her colonies as well as with other countries. From 1645 on, England passed the Navigation Acts whose aim it was to restrict colonial commerce with all save the mother country. All colonial commerce was to be carried in English ships. The Act of 1660 also listed various colonial items which could be exported only to England or her possessions. More rigid enforcement of these acts followed in 1696. It must be remembered, however, that England's colonies were less shackled than the colonies of other European countries, during the same period. The American colonies also enjoyed, to a large extent, a local self-government. The Navigation Acts were often harsher in theory than in practice.

The trouble began when England found herself with a debt of £140,000,000 as a result of the French and Indian War. English statesmen argued that, as their debt was largely incurred by a war with the French for the benefit of the American colonies, the colonists should pay a large portion of it.

Parliament Purposed a System of Taxation. The Result.—Parliament, to utilize this justification, began in 1764 to enact a series of laws with the aim of raising funds by colonial taxation. The first of these was the Sugar Act, which imposed import duties on coffee, wines, silks, and indigo; and raised those on molasses and sugar. In 1765 the Stamp Act was passed, providing for the use by the colonists of stamped paper for all legal documents. Opposition to these acts was widespread. Colonial leaders brought forth the idea that the colonists were not represented in Parliament, hence, according to Magna Charta, Parliament had no right to tax them. Next followed the Townsend Acts and the tea tax. Parliament, in view of the adverse public comment, repealed all regulative acts save the tax on tea. But it was too late. The Boston Tea Party had taken place, and the Battle of Lexington occurred shortly after. This was in 1775. The War of the Revolution, with its hardships, heroic leaders, and remarkable result, lasted until 1783, when England recognized the independence of her former colonies.

The Financial Difficulties of the New Republic.—The Continental Congress faced one of its chief problems in the question of financing the war. Arms, supplies, and money were the necessities of the colonists. Money was raised by loans from Americans and also from sympathetic foreigners. These sources were insufficient to meet the needs of thirteen colonies. The Congress lacked the power to levy any taxes, which had been so unpopular with Americans. As a result, in June, 1775, bills of credit were issued to the sum of $2,000,000 based upon state credit, to be redeemed in silver after 1779. By that time, the country was flooded with paper to the amount of $241,000,-000 and each dollar was worth between two and three cents in coin. The Continental paper was not secured by any reserve, and hence depreciated rapidly. By 1781 the Continental money had become valueless, and was no longer accepted in circulation.

The Expansion of the Country.—The territorial extent of the former English colonies was almost identical in area with the United States east of the Mississippi River, except Florida. This gave to the new country territory vast in extent, resources, and potentialities. Only the eastern fringe was populated, and the interior was an undeveloped wilderness. The development of this soon became a national undertaking.

The pioneers for the next fifty years pushed back the frontier little by little. The Middle West, then the Mississippi Valley, and later the Pacific slope saw the inrush of the settlers. As the territorial expansion increased, the need for transportation made itself felt. The story of the American railroads is a full chapter in the industrial history of the country. Also, in order to develop the new lands, American ingenuity produced machinery, and soon America was using more types of machines than almost any European country.

Social attitudes and ideals, too, were developing, and the characteristic American individualism in thought and business had its root in the pioneering days. The government aided the individualistic spirit by lavishly dispensing monopolies, privileges, and land grants. The path from worker to manager had, from the very extent of the work to be done, been kept open. Democracy in industry was the concomitant of democracy in government. As a result, classes did not crystallize as they did in Europe. These conditions, too, made possible the intelligent participation of the public in governmental affairs.

The First Census, 1790.—The first census was taken in 1790. The total population was 3,927,214, of which most was centered along the Atlantic coast. Virginia led with 747,610, Pennsylvania had 434,373, North Carolina 393,751, Massachusetts 378,787, and New York 340,120. Only 3.3 per cent lived in cities or towns of 8,000 or more inhabitants. It has been estimated that 95 per cent of the people lived by farming. The townspeople were engaged in commercial and mercantile pursuits in most cases.[1]

England's Policy Delayed the Industrial Revolution in the United States.—Although Great Britain was forced to recognize American independence, she attempted to retain trade with the United States for herself. She still wished to get raw materials from America and to sell her manufactured products there. English merchants often sold at a loss in order to keep the American trade. England's policy delayed the Industrial Revolution of the United States until after the year 1812, when, because of war with England, the Americans found it necessary to rely on domestic manufactures. Until that time Britain had attempted to discourage American commerce.

Industrial Growth After 1820.—Beginning about 1820, the growth of manufactures within a few generations made the United States pre-eminently the industrial nation of the world. First slowly, then, after the Civil War, more rapidly, the evolution of American manufacturing proceeded on a scale even greater than that of England's Industrial Revolution.

This growth was a natural development. The nation, it has been shown, possessed vast supplies of raw materials. The people had been forced to manufacture when outside influences

[1] Figures given in Cowdrick: "Industrial History of the United States."

had made importation impossible. Now, with the secrets of the manufacture of factory machinery known, a protective tariff in operation, and unfriendly relations with England, domestic manufacturing received an impetus.

The Beginnings of the Factory System in America.—It is interesting to note the beginnings of factory manufacture in America. England jealously guarded the secrets of spinning and weaving machinery that had caused her Industrial Revolution. In 1774 Parliament passed a law prohibiting the exportation of any tools or machinery used in cloth making, or any plans or drawings of these machines. Later it was made a crime to induce English factory workers to leave the country, and an attempt was made to prevent emigration of any who had knowledge of the construction of factory machinery.

This delayed the American development, but only for a short while. A spinning machine was set up by Christopher Tully at Philadelphia in 1775. Beverly, Massachusetts, had a cotton factory by 1787. All knew how Samuel Slater, unable to carry away plans from England on account of the rigid enforcement of Parliament's law, kept in his memory the plans for spinning and weaving machinery. He set up a factory at Pawtucket, Rhode Island, in 1789. With the secrets once out, the Americans not only constructed machinery of their own, but incorporated many improvements in its construction. By 1820 America was well advanced in manufacturing under modern methods and on a scale of real size.

The Textile Industries Developed Early.—As in England, the textile industry was the first to be adapted to factory machinery. The abundance of raw material and water-power soon pushed the cotton industry to the fore. The Whitney cotton-gin, the cheap slave labor, and the demand for cloth all aided this industry. In 1803 the United States had four cotton factories; by 1808 there were fifteen, with 8,000 spindles. In 1815 there were 500,000 spindles. In 1800, 500 bales of cotton were used; while in 1815, 90,000. Francis Lowell introduced the power loom at Waltham, Massachusetts, in 1814, giving another impetus to the industry. The wool industry, too, had developed.

This industry differed from the cotton, in that it suffered from a lack of raw material and insufficient tariff protection. About 50,000 workers were employed by the woolen industries in 1815, with a product valued at $19,000,000.[2]

The Metal Industries.—The iron and steel industry is of more recent development. Its needs, perhaps, were not so pressing as those for textiles. About 1800 eastern Pennsylvania produced some iron. This was due to the fact that ore and wood, for smelting, were plentiful there. Pittsburgh had a small furnace in 1792, built by George Anchutz, which, after two years of operation, closed down for lack of ore. By 1829, however, that city, now the center of the iron and steel industry, had 8 rolling mills and 9 foundries. Cowdrick, page 76, estimates the pig iron production of the United States then as being about 100,000 tons per year. The fuel for the iron furnaces was wood charcoal. The limitless forests of the United States, it was thought, could furnish an endless supply of cheap fuel. Not until 1840 was anthracite used in the iron industries, although its proximity to the furnaces had been known some time.

The First Federal Report on Manufacturing—1791.—The first governmental report on the manufacturing industries was made by Alexander Hamilton in 1791. He noted the following manufacturing industries: leather, iron, tools and machinery, hats, oil, sugar, hardware, carriages, tobacco, and gunpowder. He stated that these had developed to such a degree that they could be classified as organized industries. He further noted that many industries were yet carried on as household manufactures. Another Secretary of the Treasury, Albert Gallatin, in 1809 made a report on the condition of industries. He estimated the value of the manufactured products at $120,000,000 a year.

Labor in the Early Days of the Republic.—With the development of manufactures, the industrial wage earners became an increasingly important class. The factories, as in England, soon drew to themselves great numbers of workers. The work-

[2] Figures from Cowdrick: "Industrial History of the United States."

ers were better off than their European brothers at that time, although most historians admit that their condition was far inferior to that of present-day workers. The hours of labor extended from sunrise to sunset. Their living conditions were poor, their homes dingy, their food coarse, and their opportunities for social betterment meagre. However, they could raise themselves from this position of lowliness more easily than could English laborers, because of the great opportunities for work in the developing hinterland.

There was little organization among American workers prior to the Revolution. The early nineteenth century saw the birth of labor organizations in America. In 1803 the New York Society for Journeymen Shipwrights was formed. The New York carpenters organized in 1806, and the typesetters of the same city united in 1817.

In 1827 a strike for a ten hour day in the carpentry trade led to the formation of the Mechanics' Union Trade Association. This organization the following year named candidates for public office and several of them were elected.

Ezekiel Williams ran for governor of New York in 1830, having been nominated by a working men's convention at Syracuse.

By 1832 there was a general Trade Union of New York City, made up of various local organizations.

However, the unions of this period gave little indication of the power they were to wield in later days.

The Development of the West.—The next chapter in the industrial evolution of the United States concerns itself with the opening of the West. The events and the results of the development of this great section of the country are of major importance in the history of America. Many of the roots of America's industrial supremacy are to be found in the story of this remarkable development. Within a few years the vast trackless wastes of forest and prairie were brought into cultivation. Forest paths and trails gave way to railroads, and scattered Indian settlements disappeared in the growth of cities. Attention is often given entirely to the new West, its growth as a source of food and raw materials, and this is often the chief interest of industrial histories. While this development was going on, a decided change took place in the old East. The eastern farmer with his non-productive land could not meet the competition of the farmer of the West with his fertile farm, hence factories took the place of farms in the East. The Atlantic States, especially the northern ones, came by the end of a few years to be the center of manufacturing, commerce, and finance. They depended for their food upon the other parts of the country.

At the close of the Revolution those states which had claims upon western lands ceded them to the central government. The Northwest Territory was created by act of Congress in 1787. This included the land between the Great Lakes and the Mississippi and Ohio Rivers. The soldiers of the Revolution received lands as payment for their services. This was the first movement to settle the new territory. Within a few years, swarms of men, women, and children poured over the mountains into the fertile fields of the Ohio Valley. Congress sold large tracts of public land to companies, who sought to dispose of it to farmers. Due to popular distaste for this system, the Government soon began to sell tracts as small as 640 acres.

The first pioneers engaged in agriculture. Trade and production were delayed by the lack of rapid transportation. The fertile soil of the Ohio valley soon produced an over-abundance of grain, live stock, hides, and furs. These products were floated down the Mississippi in a lucrative trade with New Orleans. In 1803, the Louisiana purchase added to the territory to be developed. The Lewis and Clark expedition of 1803-1806 brought back such accounts that the westward migration continued, and settlement of the Oregon territory began. Then in 1848 came the discovery of gold in California, and the migration took on the character of a rush.

Transportation.—It has been noted at various points that the lack of transportation facilities constituted a hindrance to full development. The size of the country, the distance between centers of production, and the relatively sparse population enhanced the need for a system of transportation and intercommunication which has been furnished largely by canal, road, railroad, automobile, telegraph, radio, and wireless.

In colonial times traffic between cities was carried on by horse on roads of the crudest construction. A coach journey from New York to Boston was a trip of three days. On the coast, where the largest cities were located, transportation was by water in sailing vessels. Turnpikes and improved roads came into being after 1790. The toll collected from travelers went to pay for the building and upkeep of these roads. Soon all the East was connected by a net work of these highways.

Steam Applied to River Navigation.—But the distance between points in the East and those in the new West demanded even better means of transportation. Steamboats came as a first answer to this need. In 1807 Fulton demonstrated his practical project when the "Clermont" steamed from New York to Albany in 32 hours. The general application of this was hindered by a twenty-year monopoly granted by New York to Fulton. A similar monopoly failed to tie up the Mississippi, when the Supreme Court of the United States held that the use of the rivers could not be monopolized (Gibbons v. Ogden 1824). The use of steamboat transportation rapidly spread after that. The New Orleans trade of the settlers in the Ohio-Mississippi Valley received an impetus. In 1816 the produce received at New Orleans approximated $8,000,000, 8 per cent of which originated in the upper Mississippi and Ohio Valley. In 1829 this had grown to $22,000,000; by 1840 to about $50,000,000.

The Canal-Building Era.—The era of canal building parallels that of steam navigation. The most famous canal was the Erie Canal, built during the years of 1817 to 1825. This water route offered an outlet for products within any short distance of the Great Lakes to be carried to Buffalo and thence to Albany by means of the canal and finally down the Hudson to New York City. Rates fell to about one tenth of their old cost, and the shipping of New York City increased at a rate that soon put that city at the head of the list of commercial centers of the United States. Other canals followed, prompted, no doubt, by the success of the Erie Canal.

The First Railways.—The era of canal building was stemmed, however, by the innovation of an even more efficient type of transportation. The first railroad for commercial use was started in 1828, when the Baltimore and Ohio Company built a road out of Baltimore. Thirteen miles were open in 1830, while seventy miles were in use the following year. In 1840, the railroad mileage was 2,818, by 1850, 9,021, and by 1860, 30,626. The railroads closely followed the pioneers, and helped bind the country together. They furnished the means for the interchange of products between the East and West, as well as the North and South. It is safe to say that the United States could not have developed as rapidly commercially or socially without the railroads.

Early Tariff History.—A word on governmental control of industry is pertinent at this point. In order to protect infant industries, Congress has followed the policy of protection by establishing tariffs on imported goods. The tariff of 1789 was, on account of its low rates, a tariff for revenue only. This act had within it the germ of the later tariff system. Its purpose was to protect and encourage the growing industries. After Congress convened in 1789, James Madison proposed a resolution calling for the adoption of imposts. He stated that his purpose was to raise revenue. Mr. Fitz Simmons of Pennsylvania, who recognized the difficulty of changing a law that was already enacted, proposed the system of protection of his state. It contained a list of articles which had been taxed by the colonies. The Fitz Simmons plan was adopted, and from that time protection has maintained its place in American politics.

From 1810 to 1816 a greater trend towards protection is noted. The really first protective tariff was the tariff of 1816. The highest rate was 20 per cent; and the reason for the increase was the indebtedness of the country after the War of 1812. The birth of new industries crying for protection was another reason. The middle and north Atlantic states and the

West were the strongholds for protection. The South took its stand against protection.

From 1820 to 1837 the tariff fluctuated. In 1820 a high tariff bill was proposed but failed to pass. Like bills were proposed in 1821 and 1822. They, too, led to no legislation, and it appeared that the protective movement had lost its impetus. Then came the presidential election of 1824, which was the first and most direct fruit of the early protective movement. Party lines were shattered as the bill of 1824 went through. This tariff law provided for increased duties on iron, lead, wool, and hemp. The average increase was from 25 per cent to 33 1/3 per cent.

After 1824 there was another lull in the agitation for protection. Business and trade were buoyant in 1825. Then came the bill of 1833, called the Compromise Act. This satisfied the public demand for lower tariff rates providing for a general and gradual reduction. In 1842 a reduction of 20 per cent was reached just two months before the passage of the tariff of 1842, which was decidedly protective.

In the years 1840–1860 there was great vacillation in tariff policy and also great fluctuation in trade and industry. Low tariffs succeeded high tariffs, and vice versa. The tariff of 1842, passed by the Whigs as a party measure, was a protective tariff. The law remained in force four years to be followed by the act of 1846, passed by the Democrats; and it was a moderation in the application of protection. This act remained on the books until 1857, when a further reduction in duties was authorized. Revenue was plentiful in 1857, and this was the cause of the reduction. This was the first tariff since 1816 which was not affected by politics. All parties agreed to reduction due to general prosperity. The act of 1857 remained in force until the war period of 1860–1870.

Industrial Progress from 1820–1860.—After 1820 home manufacturing was firmly established and a general growth took place. The increased demand for manufactured goods, spread over larger areas, created markets and increased profits and production. A series of mechanical inventions, too, took place. Agricultural implements were crude and had come down from antiquity. The mechanical reaper of Cyrus McCormick in 1835 was a first step in mechanical farming. Other inventions and applications also aided the farming activities.

Elias Howe in 1846 invented a sewing machine. This invention proved not only a household aid, but also a necessity in the tailoring industry. The principle of the sewing machine was soon applied to the leather industry and the boot and shoe industry.

The publishing industries also improved, due to new devices and improvements on the printing press.

The magnetic telegraph also had a potent influence on the evolution of manufactures. In 1843, Professor Morse constructed a telegraph line from Washington to Baltimore. A little later came the first Atlantic cable.

Coal was adopted more generally as a fuel after 1830. This brought into being the coal industry, and provided for the increased use of steam power. Anthracite had been known in Pennsylvania and bituminous had been known in Virginia and Ohio and Illinois. The coal fields of Pennsylvania were not worked before 1820 due to poor transportation facilities and also to the popular belief that coal would not burn. The use of coal gas for lighting also called forth a demand for coal. The gradual disappearance of the forest also built up a market for coal as a domestic fuel. Coal mining in its present form, however, was not developed until after the Civil War.

Textiles, iron, flour, and meal were the chief products of the period from 1820 to 1860. The manufacture of lumber, boots, shoes, clothing, machinery, carriages, distilled and malt liquors, tobacco products, paper, soap, agricultural implements, and marble and other stone products also found place in the list of important industries.

The manufacture of cotton took the lead in the nation's factory enterprises. In 1830 the value of cotton manufactures was, as reported by the census, $22,534,815. In 1860 it had increased to $115,681,774.

The growth of the wool industry, hindered by the lack of raw material, did not develop as rapidly. Wool goods amounted in value to $4,413,068 in the 1820 census, while by 1860 they had reached a valuation of $73,454,000.

The growth of railroads and the development of machine manufacturing increased the demand for iron. Between 1820 and 1860 the iron industry kept up with this augmented demand. By 1850 the Lake Superior ore was being worked, while by 1860 the Michigan output was 114,000 tons. Pennsylvania led with a total of 1,706,000 tons. By 1850 coke was used as the reducing fuel. Steel was as yet only a small part of the iron industries.

The period just considered, 1820–1860, saw the transformation of the United States from an agricultural to an industrial nation. By 1840 the large industrial organization was quickly replacing the small factory. A labor class of factory workers now existed, whose difficulties and treatment constituted a "labor problem." Noteworthy changes occurred in the industrial system. A capitalistic class had grown up which operated and directed the nation's wholesale business. Selling direct to the consumer had given way to selling to a middleman. Labor exploitation was a common practice. Hours of labor were long, wages unsteady; prison labor, sweat shops, and other undesirable conditions made it inevitable that labor should, as in England, organize for protection.

Labor Activities, 1820–1860.—About 1827 the influence of labor organization began to be felt. These organizations joined in an effort to secure, through legislation, certain objectives. They wanted a ten hour day, abolition of imprisonment for debt, restriction of child labor, free public education, and other aims, many of which have since been obtained.

Workingmen's parties were organized in many cities for the purpose of electing selected candidates for office. After 1830 this political activity passed, although it left its stamp on social legislation for the next decade.

In 1836 Philadelphia had fifty-three trade unions, while New York had fifty-two. In 1834, 1835, and 1836 attempts to form a national federation failed.

The panic of 1837 left a depressive effect on the country's

business. The unions could not hold their members nor collect dues from the hundreds of thousands out of work. This was a period when many of the unions disappeared.

The increased immigration following 1847 caused the American workmen to fear competition of European workers. The Native American Party of 1847 gave expression to this dread. Immigration had materially increased due to several fundamental causes, the more potent of which were the failure of the potato crop of Ireland in 1846, the political upheaval in Europe in 1848, and the discovery of gold in California in 1848.

Business revived after 1837. Labor turned from its political activity of the depression period to activity in normal industrial pursuits. The period of 1850–1857 saw more strikes than any previous period. Then followed the panic of 1857. Banks failed, business closed down, and labor was unemployed. Unions again had to fight for their lives. The Civil War came as a climax of this era.

The number of people in industrial pursuits during the period 1820–1860 was only a fraction of the population. An even smaller number of those belonged to labor organizations. All laborers shared in the general prosperity or depression of the period. The census of 1850 gives the number of those employed in industrial production as 957,059. The yearly wage was approximately $250. Increasing numbers of women were now being employed in industry. The early prejudice about the woman's place being in the home was being overcome. Twenty-three per cent of the employees in 1850 were women. Children found their especial place in the textile mills.

Another reference to the immigration of the period must be made. The countries which supplied the immigrants were largely England, Ireland and Germany. In times of prosperity the immigrants made a needed addition to the working force of the country; in times of depression every immigrant was looked upon as an intruder who menaced the work of the natives. The immigrants who came largely from English-speaking countries created no problem similar to that created by the southeastern Europeans of today. The newcomers were easily assimilated and incorporated in the state with ease and benefit. The immigrants settled for the most part in the North and West, avoiding the South where slavery blighted free labor and drove it from competition.

The South Became a One-Product Section.—Despite the development of industry in the North, the South remained agricultural. Furthermore, the methods of Revolutionary days persisted with little change below the Mason-Dixon line. Cotton was the one product of the South, and this crop was raised and harvested by negro slaves.

Labor in the South.—In colonial days negro slavery was more or less general throughout the colonies. In the North, however, the climate was unfavorable for the African. Also, slave labor did not lend itself to the factory system and its intricate machinery. In Virginia the slaves proved amply capable in the tobacco fields, and in the rice fields of the Carolinas they carried on the work most efficiently. At the time of the Revolution there is evidence that some thought was being given to the question of the ethics of slavery. Virginia, as well as New England, was the scene of an anti-slavery agitation.

Slavery Firmly Established in the South.—As cotton became more and more the fundamental product of the South, however, the anti-slavery sentiment disappeared. By 1807, 80,000,000 pounds of cotton were produced in the southern states. England supplied a steady export market, while the markets of the northern states provided a domestic one. In 1830, the cotton crop had grown to 350,000,000 pounds. This increase called for a great number of workers. White hand labor was not sufficient to meet this need. The supplying of slaves then proved a profitable occupation. In the advent of many slaves free white labor withdrew. The South, from an economic standpoint, then rationalized the institution of slavery.

In 1808 the importation of slaves was forbidden by Federal statute. This law was easily broken by smuggling. The profits proved large enough, in many cases, to defray the price of fines, if the smuggler were caught. However, by 1820 slave trading was classed with piracy as a crime punishable by the death penalty. This practically cut off the outside supply of slaves.

The natural increase of the negroes in the country provided for the perpetuation of the system. The 1830 census showed the number of slaves to be 2,009,043; in 1860 this number had grown to be 3,953,760. This great increase, however, fell far short of the demand, and consequently the price of a slave rose. In 1790 an average male slave brought $200 in the auction market; by 1860 the price hovered around $2000.

Slave labor was crude labor. The South did not build factories, encourage manufacturing, or develop scientific agriculture. Reliance was placed on the North and on Europe for manufactured goods. Slavery kept the South a one-product region. Truly, "cotton was king". The middle-class whites were largely driven out of the South, for they could not compete with slave labor.

The returns from cotton culture proved so profitable that its spread to the new Southwest began as soon as that territory was opened for settlement. Of course, slavery went hand in hand with the cotton culture. The one crop system was exhausting the soil of the old South, while the new West offered a virgin field for development. Tennessee, Louisiana, Mississippi, and Alabama were soon raising cotton. In 1835 these four states raised nearly two thirds of the total annual crop.

The Missouri Compromise, 1821.—When Missouri was ready for statehood, it applied for admission into the Union as a slave state. According to previous custom, the slave population was counted in allotting representatives to the state for seats in the House of Representatives in Washington. The usage at that time was to count three fifths of the slaves. The North took alarm, thinking that all the Louisiana purchase would be cut up into slave states and thus give the South control of the House as well as of the Senate. After some time and much debate, a compromise was agreed upon, the famous Missouri Compromise of 1821. Missouri was allowed to enter the Union as a slave state and, at the same time, Maine was to enter as a free state. The compromise also provided that no more slave states should be created in the Louisiana territory north of Arkansas.

The Annexation of Texas, 1845.—In 1836 Texas freed herself from Mexican control. Texas at this time contained a large American population. The question of annexation of Texas by the United States came rapidly to the front. This talk created a furor in the North, for Texas would be, if annexed, a slave state. The campaign of 1844 was carried on by the Democrats on a platform for annexation. The plan was to allow Texas, and other states to be cut from Texas, to choose, by vote of their people, between free and slave statehood. Texas was admitted to the Union in Polk's administration. This event led to the war with Mexico in 1846. As a result of this struggle, the United States added, besides Texas, New Mexico, Arizona, California, Nevada, Utah, and a part of Colorado.

The Acquisition of Oregon, 1846.—In 1846 the United States acquired possession of the Oregon country. The territory included the present states of Oregon, Washington, Idaho, and parts of Montana and Wyoming.

The Discovery of Gold in California, 1848.—Soon after 1848 gold was discovered in California. The gold rush was on. The population of California grew so that statehood was soon demanded. The people were largely from free states and slave labor could not advantageously be employed. To admit California as a free state would destroy the traditional balance of power. Some people wished to continue the line 36°30′ to the Pacific coast and thus perpetuate the Missouri Compromise. The radical advocates of slavery questioned the right of Congress to prohibit slavery in any of the new dominions. The abolitionists wished to prohibit slavery in all the new territories. Henry Clay advocated and had adopted the Compromise of 1850. This was just another temporary measure to delay a final decision. It allowed California to enter as a free state, while Utah and New Mexico were organized as territories with the provision that the people thereof should decide upon the question of slavery. A further provision was a prohibition of slavery in the District of Columbia and a strict Fugitive Slave law. Within four years the question was again ablaze. The Kansas-Nebraska Act of 1854 left it to the people of those territories to decide upon slavery. Civil disorders in Kansas between free soil and slavery supporters grew from year to year. In 1860 Abraham Lincoln was elected President. The South formally withdrew from the Union to set up the Confederacy.

The Civil War.—A four-year war of untold suffering and misery was required to settle the question of slavery. The war proved that the South with its one product had not such resources to rely upon as the North had with its manifold types of industrial manufactures.

The cost of the war to the North was about $2,000,000 a day. The story of how this tremendous amount was met is a whole chapter in the financial history of the country. In brief, the Government met the bill by increased tariff rates, by the issue of bonds, by internal revenue taxes, and by the issuance of Treasury notes which circulated like money.

The Civil War and Its Industrial Effects.—The war played havoc with the normal industrial evolution of the country. Its influence upon trade, industry, and labor is incalculable. In the North, those businesses which carried on trade with the South either failed in the early days of the war, or else diverted their business to other channels. Foreign commerce with the North, too, was disturbed. High tariffs and duties discouraged foreign merchants. The high tariff shut out European goods, and hence the European could not afford to make purchases in the North. The privateers of the South also preyed upon northern commerce. America's place as supreme ocean carrier was lost during the war and her place was quickly filled by Great Britain.

Manufacturing, however, received an impetus in the North. The demand for clothing, food, weapons, and railroad equipment gave industry all that was necessary for a big development. Then, too, the tariff well protected the manufactures. In 1860 the census showed the invested capital in manufacturing plants to be $1,009,885,715 and the number of employees, 1,311,248.

In 1870 figures for the same items were $2,118,208,769 and $2,053,996.[3]

Agriculture in the North had to become more productive to meet the increased need. The new machinery enabled farmers to feed the armies at the front as well as the people at home.

Industry in the South During the War.—Many industrial histories neglect to survey the conditions in the South during the war. Cotton proved to be a worthless product to the Confederacy. The blockade cut off the foreign market for this product, and the South had not developed textile factories to manufacture cloth. As a result, cotton rotted upon the wharves, and the cotton planters were thrown into bankruptcy. The pre-war profits from cotton culture had precluded the development of food raising, and the wartime South had difficulty in securing a food supply, with European and northern food cut off.

An attempt was made to build up a manufacturing system in the South. War material, clothing, and machinery were produced in hastily erected factories. Trade activities suffered, as could be expected, due to the cotton collapse, the deadly blockade, and the financial demoralization. The realization of the loss by the freeing of the slaves caused a further halt in business toward the end of the war.

Labor, in the North, During the War.—Abundance of employment, in the North, gave labor a favorable market during the war. The factories called continuously for men. The fact that so many went to war left a diminished supply, which could demand favorable terms. Trade unions attempted to increase wages in proportion to the increased cost of living. By the close of the war there were many nationally-organized unions. Labor met a snag, however, in the Alien Contract Immigration Law of 1864 which permitted employers to import laborers under contract to work out their passage money. Immigration increased, and the American laborer shouted loudly his disapproval of competing with Europeans and their low standards of living. A few years later pressure from labor organizations caused the repeal of the law.

[3] Figures from Cowdrick: "Industrial History of the United States."

At the Close of the Civil War, the Factory System Was Firmly Established in the United States.—The close of the Civil War marks the date of a new industrial era in the history of the United States. The Industrial Revolution with its factory was firmly established, and capital established the manufacturing and commercial supremacy of the nation.

The financial reconstruction was a fundamental problem of the after-the-war period. Slavery was finally abolished, and the reunited country turned to cope with the new problems of a dominant industrial power in politics, industry, finance, and society.

The Panic of 1873.—Speculation, particularly in gold, and an attempt to corner that market gave rise to the panic of 1873. Reckless speculation of the type of the Gould and Fisk practices in the Erie Railroad were potent causes of this financial crisis. The high protective tariffs and the rapid construction of railroads, often built on shady systems of credits and in territories where immediate returns could only be low, led to feverish activity in business. Railroads grew by 33,000 miles in the interval 1865–1873. Stocks of railroad companies and of all industrial enterprises soared. Cities indebted themselves by contracting for expensive improvements. Even small farmers invested more than they could afford in new lands.

European contraction led American bankers to call in their loans and raise the call money rate. In September, 1873, Jay Cooke & Company failed. This company held the obligations of the Northern Pacific Railroad Company. This was the signal; the financial structure of the country crumbled. The New York Stock Exchange closed its doors for eight days. Although the panic soon passed over, business remained in a stagnant slump for at least five years. Not until 1878 did the country again recover industrially and commercially.

Western Development After the Civil War.—After this period of depression, the natural field for development was the West. Settlement of the West had been going on for some time, aided often by such attractions as gold discovery and the chance for more profits in farming. When the war came, thousands of westerners swelled the ranks of the warriors, only to leave those at home in need of food. To secure this, improved methods of farming, mechanical devices, and the cultivation of more land were taken up. Returning soldiers who found themselves out of the swing of the social life of the East migrated to the new West. The mines of Colorado and California, together with the prairies of Kansas and Nebraska, offered harbor to those dissatisfied thousands.

In 1862 the Homestead Law was passed. This provided that any person over twenty-one who was the head of a family and who was a citizen or had filed his intention to become a citizen, could enter upon a tract of 160 acres of unappropriated public land. Upon five years of cultivation of this land, title of ownership was given. This law differed radically from previous encouragements for western settlement in that it gave, rather than sold, the public domain. The passage of this law created much discussion as to its wisdom. Its advocates declared that it encouraged farming and western settlement and increased the food supply. They inferred that it caused stability by providing opportunities for home owners to make decent livings. The opponents of this measure lamented the fact that the major portion of the desirable public domain was squandered in a single generation. They decried the frauds connected with the dispensation of the grants. Sixty-five million acres were granted in the years between 1862 and 1880. The frontier was pushed back to the Pacific.

The prairies of the West had lent themselves to the cattle-raising industry. The coming of the homesteaders threatened to stamp out the industry of the open-range cattle men. The warfare between the cattle men and the homesteaders was bitter and lasting. In the end, intensive agriculture triumphed over the extensive cattle raising. When the homesteads finally cut the range lands down to a minimum, the cattle men took to intensive cattle raising and the improvement of cattle needs.

Agriculture in the Modern Era.—Agriculture was intensive as well as extensive at this time. The improved farm land amounted to 163,000,000 acres in 1860; in 1880, to 284,000,000;

THE ERA OF RAILWAY BUILDING

in 1910 to 478,451,750; and 1920 to 503,000,000.[4] The grain area of the north central states was developing, with wheat and corn as the chief products. Threshers, tractors, harvesters, cultivators, planters, and other mechanical implements were applied to farming and made farming possible on a large scale. Irrigation in many places had to be provided in order to give fertility to the soil. Lack of railroad facilities and high freight rates caused great distress to the farmers. Farmers from 1870 to 1910 were getting increasing returns year by year. The World War brought still greater demands for food products.

Reconstruction in the South.—The South, wrecked by the war, staggered through the reconstruction period. Cotton culture on the basis of slave labor was broken up. The planters now set to work to establish the cotton industry on the basis of free negro labor. The end of the Civil War saw a boom market for cotton. The war cut off the supply of southern cotton for four years; this raised the price of cotton from 30 to 43 cents a pound. The increased cost of labor made the plantation system inefficient, and the large tracts of pre-war days were divided into smaller sections. These were often worked by freed slaves who paid rent in the form of a percentage of the crop raised thereon. The South, since 1870, has diversified its agricultural activities, has erected factories, and has developed industrial and mining districts.

The recent problems of the South have been the destruction of the cotton crops by the boll weevil and the migration of negroes to the higher-paying industries of the North.

The Era of Railway Building.—National development, especially in the region west of the Mississippi, called forth the building of an extensive net work of railroads. The history of the West from 1860 was the history of the efforts of the railroads to parallel the development of the region. Railroad building prior to the Civil War was slow. In 1860 there were but 30,626 miles of track laid. Of this mileage, 30 per cent was in the South, and the greater part of the remainder was east of the Mississippi. The Civil War, as could be expected, inter-

[4] Figures from Cowdrick: "Industrial History of the United States."

rupted the natural continuation of railroad building. Soon after the outbreak of the War, demands in addition to the normal industrial and commercial needs brought about the great railroad building era. By 1870 there were 52,914 miles of track; by 1880, 93,296, and by 1890 the figure reached 163,597. The 1920 figure was 253,152.

The gold seekers of 1848 demanded supplies from the East, and sought a means to ship their products back there.

Overland wagon travel over rough trails was long, laborious, and not dependable. In 1860 the discussion of the building of a coast-to-coast railroad came up. The obstacles to the proposal lay in the fact that the projected line would run through miles of the desert with no inter-station traffic to offset construction costs. It was argued, however, that such a road was a public benefit, and would be a means of binding the East to the West. The North felt that, as the West had given aid during the Civil War to the anti-slavery cause, it would be fitting to reward that section by affording it transportation facilities.

In 1862 Congress granted aid to the Union Pacific and Central Pacific railroads for the construction of a connecting line. The Union Pacific extended as far west as Omaha, and the Central Pacific had a short line in California. With an increased government subsidy in cash and lands, the companies rushed construction. The two roads joined at Ogden, Utah, in May of 1869.

After one company had built a road across the country, others soon followed. The Northern Pacific received grants of land and completed a road in 1883.

The period of governmental aid was followed by one of suspicion and regulation. Financial difficulties, frauds, watered-stock exploitations, and extravagances caused higher rates. The public objected, and the railroads and the public broke friendship. The blame rested partly with the railroad managers and partly with the public.

Railroad history next tells the story of combination. In the early days the roads were short and under individual control. By 1867 Cornelius Vanderbilt had succeeded in uniting several roads into the New York Central system. Soon afterward, the New York lines controlled the Lake Shore, the Canada Southern, and the Michigan Central, thus joining New York to Chicago under a unit control. Other powerful consolidations included the Pennsylvania and the Baltimore and Ohio.

Destructive rate wars marked the early period of consolidation. The realization of the killing effect of these rate cuts led to the pooling agreements. Land served by more than one road was divided, and each company had the benefit of all freight originating in a designated district. Earnings, too, were often pooled, to be divided later in accordance with a predetermined scale.

The Inter-State Commerce Commission.—Federal railroad regulation was first applied in 1887, with the passage of the Inter-State Commerce Law. By this act, a commission of five members was established to hear complaints against railroads and to render decisions from which appeal could be made to the Supreme Court. Reasonable freight and passenger rates were prescribed by the law as well as the prohibition of discriminatory rates. The commission could demand annual reports of the railroads and other pertinent information.

Law evasions and court decisions which limited the power of the commission rendered its service almost worthless. After eleven years the commission complained of abuses which made the railroad system anything but desirable.

Recent Railroad Legislation.—Congress then proceeded to strengthen the law and establish the commission on a firm basis. The Elkins Act of 1903 provided regulations which were backed up by more severe penalties. Violation of the law, discriminatory rebates, or the charging of rates higher than those published were made offences.

The Hepburn Act of 1906 was even more strict. Its provisions applied also to express companies, sleeping-car companies, and oil and gas pipe-line companies. Finally came the Mann-Elkins Act of 1910. A commerce court was created with jurisdiction over controversies connected with railroad transportation. This court was abolished in 1913.

The need for cheap and rapid inter-urban and city transportation developed about 1880. In 1890 electric street cars began to supplant the old, slow horse cars. The superiority of this form of city rapid transit was evident, and soon its use spread universally. The application of electricity to inter-city travel, especially in the central states, is a characteristic means of present-day transportation. Motor car and truck transportation are twentieth-century modes of transportation. In 1921 it was estimated that motor trucks carried 1,430,000,000 tons. For short hauls the motor trucks have in many cases supplanted the railroad. Aeroplane transportation is now being commercially developed; although at present (1930) it is used chiefly for mail and passenger service.

The Development of the Coal and Steel Industries.—Along with the development of railroad transportation came the increased use of coal and steel. About the year 1860 the coal output was 14,000,000 tons a year. After the Civil War, in the period of rapid growth, this figure mounted. In 1870, 33,035,500 tons were produced; by 1890, 157,770,900; by 1910, 501,596,377; by 1920, 645,616,000.

As has been noted, until well into the nineteenth century wood and wood-charcoal were used as sources of heat in separating metallic iron from the other elements in the ore. In 1840 Pennsylvania had six blast furnaces using anthracite as a reducing agent. By 1854 the use of anthracite had equaled that of charcoal, and each produced 340,000 tons of iron in that year. By 1860 the use of charcoal had dwindled. In that year 500,000 tons were produced in anthracite burning furnaces, while only about half as much was produced in charcoal furnaces.

The use of anthracite as a reducing agent had marked effects on the iron industry. Furnaces of larger size were built, and the industry became localized in centers where ore and fuel could be procured in large amounts. Thus Pennsylvania had 121 furnaces using anthracite in 1856 and Maryland had 6.

Later, in many cases, coke was used instead of anthracite. The increasing cost of anthracite led to the use of the newly

THE MANUFACTURING INDUSTRIES

discovered coking bituminous coal at Connellsville, Pennsylvania. To-day coke is made in a by-product process which utilizes the tar and gas given off when the coal is burned.

Machine industry and the railroads created a great demand for iron and steel. The pig iron production for 1880 was 821,000 tons; for 1890 it was 3,385,000; in 1919, 30,543,000 tons; and in 1922, 27,219,904.

The iron of the first part of the period was largely cast iron and wrought iron. Steel was used in special cases where extreme hardness was desired, but this was rare, for the cost was very high. By 1858 Bessemer had developed his widely known steel-making process. The product of this process was found to be better, cheaper, and more quickly produced than wrought iron. This process proved useful in cases where the iron ore did not contain phosphorus, an excess of which American ores contained. These non-Bessemer ores led to the adoption of the open-hearth method of steel production.

Iron ore exists in Pennsylvania, Michigan, Ohio, New York, Massachusetts, New Jersey, and a few other states. Pittsburgh became the center of the industry. The Lake Superior ores proved to be the greatest in extent after the Civil War. Another later development was the Alabama district, which produced 4,838,900 tons of ore in 1914. In 1910 the iron ore production of the entire United States was 56,889,734 tons (gross) and in 1920, 67,604,465.

The first steel centers were the places where ore, fuel, and flux existed together. Now, due to cheap lake transportation, centers are developing around Chicago and Cleveland. The Alabama district, too, has the benefit of proximity of coal and iron ore.

The Growth of the Manufacturing Industries.—From the last two decades of the nineteenth century to the present the value of manufactured products has exceeded that of agricultural products in the United States. The factory system left its stamp permanently upon the country after the Civil War.

The causes for the tremendous industrial growth are manifold. The abundance of raw materials, power, and transportation facilities and a broad domestic market are some of those suggested. The wealth of the country in general, as well as the protective tariff, can also be included.

A few figures will plainly show the growth of manufactures in the past half century. In 1870 there were 252,148 establishments, while the number of workers was 2,053,996. The products were valued at $3,385,860,354. In 1890 there were 355,405 establishments and a working force of 4,251,535. The products now had risen to $9,372,378,843 in value. In 1920 there were 290,105 establishments with a total force of 9,096,372 workers and a production valuation of $44,569,593,771.

Machinery and manufactured power mark the modern applications of science to production. Capital and machinery also show a tendency to concentration. The small shop has been replaced in many cases by the factory. Modern management and efficiency call for consolidations and concentrations for maximum productive power. Localization of industries has characterized modern industry also. Paterson is a silk center; Detroit an automobile center; Chicago a meat-packing one, and so on. Tradition, nearness to raw materials, nearness to markets, nearness to a labor supply, nearness to power and transportation are a few of the reasons that account for centralization.

The Modern Labor Problem.—The complexity of the modern industrial system is further increased by the labor problem. Some industries rely, to a large extent, upon immigrant labor. The steel industry and the clothing industry are two self-evident examples.

Conservation—A Desired Ideal.—Another, and promising, tendency of twentieth-century industry is the practical application of the principles and ideals of conservation. The utilization of by-products, the elimination of waste, and the development of substitutes for scarce products are a few common practices that reflect conservation.

Other Industries.—Among other industries, the textile, the clothing, the meat-packing, the food machinery, and the automobile industries are typical examples of America's development of the factory system, and indicate the extent of the Industrial Revolution in the United States.

The Organizations of Workers.—The growth of the factory system, as has been noted, created a distinct class of industrial wage earners. It has also been noted that, as their numbers increased, organizations of workers had reached a considerable development by the time of the Civil War.

By 1880 the percentage of agricultural workers had decreased from a high majority to 44 per cent of the population. By 1920 only 26 per cent were engaged in agriculture, while 38 per cent were in manufacturing and transportation.[5]

With the advent of the large factory it became increasingly difficult for the individual worker to muster sufficient capital to set up an independent shop. It also became more difficult for the industrial worker to change from one type of work to another without the loss in time and money necessitated by his having to learn new methods, procedures, and processes. The intense specialization and extreme division of labor required by efficient management tends to make the individual worker narrow and dependent upon one type of work.

Wages, however, have gone up steadily, despite the setbacks due to depressions and the readjustments called forth by the variable purchasing powers of the dollar. As the economist would have it, "Real wages have tended to increase." The power of labor organizations, the greater spirit of humanitarianism, and increasing production are some of the causes of this upward trend. Industry has yielded greater returns, and labor is sharing therein. Just how and in what proportions this increase should be divided is not settled. But the interested parties, capital, labor, management, and the consuming public will undoubtedly be able to solve this problem in the future.

The Shorter Day in Industry.—The story of the evolution of the shorter working day is a problem of interest and importance. Early labor unions sought to establish the ten-hour day. The success of this movement had netted, as a result, a working day of eleven hours by 1860. In 1880 the average

[5] Cowdrick: "Industrial History of the United States." Page 251.

working-day was about ten hours. In 1905, this had shrunk to nine and one-half hours. Since the World War the eight-hour day has been established almost universally. The introduction of the eight-hour day in the continuous-process steel industry has proved that such a reduction of hours in no great way affects production. (For further discussion of the hours of labor see William P. Sears Jr.'s "The Shorter Day in Industry," New York University—1924. College of Engineering thesis.)

Legislation for the Benefit of the Workers.—Legislation, too, has been an agency in improving the condition of the workingman. In 1866 Massachusetts enacted a law whose provisions aimed to protect children in industry. Soon after, that state had also provided a bureau of labor, and had limited the working-day for women and children to ten hours. A commission also was empowered to inspect and determine the conditions under which the workers were employed. Since then the Federal Government and the state governments have provided for the improvement of the worker through labor legislation. Women and children are given special protection in industry, their hours are determined, their working conditions are controlled, and their health is protected. The labor of men in hazardous and unhealthful occupations, too, is cared for by the Government. Some states provide for factory inspection and minimum wage scales and hours of work. Compulsory arbitration of disputes between employers and employees or public investigations of those disputes is provided by certain states. The Kansas Industrial Court, established in 1920, is an example of such a permanent arbitration board. Compensation for injuries received at work is another form of legal safeguard to labor.

The results of the improved conditions of labor are immeasurable. Greater happiness of the worker and his family is one; better living conditions another. Higher wages, also, have made the worker a better and wiser consumer. Educational and social advantages have been placed within the reach of the workingman for both himself and his family. Society,

THE AMERICAN FEDERATION OF LABOR 49

too, has been a gainer in having for its members a happier and more contented people.

The organization of labor pushed steadily onward after the Civil War. Times of prosperity are times when the ranks of the labor unions are swelled. At the close of the World War, Cowdrick[6] estimated the membership to be more than 5,000,000 men and women.

The National Labor Union.—The National Union, established in 1866, was an attempt to federate city labor assemblies into a national body. Its work aided the passage of an act in 1868 by Congress which limited the working-day to eight hours for government employees. Its efforts, also, may have influenced the establishment of the United States Bureau of Labor. By 1872 it had almost ceased to exist, for it had deviated from its original course and become involved in political and reform movements. Those members, and they were a large number, who wished to restrict the activities of the organization to industrial aims, fell away.

The Knights of Labor.—The Knights of Labor was organized at Philadelphia in 1869. This was an idealistic organization, in many ways secret in character, which sought to secure certain benefits for labor. The referendum, the creation of a Federal labor bureau, the single tax, health protection for workers, the income tax, compensation for industrial injuries, and compulsory arbitration were a few of its planks. Public suspicion of its secrecy, however, together with internal factional fights, permitted it only a short existence, and by 1898 the organization was no more.

The American Federation of Labor.—A rival, and somewhat different, type of organization hastened the death of the Knights of Labor and supplanted it. This was the American Federation of Labor. A group of seceding members of the Knights of Labor met in 1881 and organized the Knights of Industry. In 1881 this society called a meeting at which 107 delegates represented 262,000 workers. The convention created the Federation of Organized Trades and Labor Unions of the

[6] Page 260.

United States and Canada. With slight changes this became the American Federation of Labor in 1886. This body is a loose federation of self-governing unions. Authority to call or end a strike is not given to it. In 1899 the Federation had about 300,000 members; by 1904, 1,650,000; and by 1920, over 3,000,000.

The I. W. W.—The Industrial Workers of the World, the I. W. W., is the most radical and revolutionary type of labor organization of any importance in the United States. It is a union of all working men, and has appealed especially to the unskilled workers. This organization was founded in 1905 at Chicago. It reached its greatest strength perhaps in 1912, when it led the strike of textile workers in Massachusetts. During the World War its leaders were prominent in anti-military activities.

The relations of employer and employee have significant influences upon the social, industrial, and educational development of the country. The prosperity of America rests on a broad understanding by capital and labor of their common difficulties. The past antagonism of employers and employees has resulted in unhappiness, waste, and discontent. All parties concerned have been losers from the past wars of labor and employers. The strikes have been the most bitter method of settling industrial disputes.

Strikes.—Among the more important strikes in the last fifty years have been the railroad strike of 1877, the Homestead strike in 1892, the anthracite coal strike of 1902, the steel strike of 1919, and the coal strike of 1922. No restatement of the causes and results of these strikes is necessary here. It is only fair to say, however, that the peaceful settlement of industrial disputes is a more efficient and more socially desirable method than the methods of the strike and the lockout.

Industrial Cooperation.—Amicable relations between worker and employer have brought about arbitration of industrial disputes, public labor tribunals, cooperation, and profit-sharing and stock-ownership systems.

Big Business Evolves.—The last decade of the nineteenth century and the opening of the twentieth saw the evolution of "big business" and the concentration of capital. The corporation, with its advantages and disadvantages, is characteristically modern. The Standard Oil Company, the General Electric Company, the New York Central Railroad, and hosts of other large corporations are examples of the trend toward consolidation. Those institutions which sometimes border upon monopolies can often be regulated by governmental supervision of their activities.

Recent Tariff History.—A brief review of tariff legislation from 1860 to 1930 is in order at this point. The act of 1857 remained in force until the beginning of the war period of 1860–1870. The financial policies of the United States were revolutionized by the Civil War. The war tariffs soared above all previous tariffs. Congress passed, together with the tariffs, great internal revenue measures. In 1864 a war-time tariff act was passed. This act raised the duties greatly and indiscriminately. From an average of 35 per cent, duties were raised to 48 per cent. The advocates of this act claimed that American manufacturers needed the aid of this act to keep going and to pay the heavy internal revenue taxes. After the war Congress failed to reduce the tariff. The internal revenue bills were repealed, but the tariff remained on its high perch. In 1867 a bill was introduced which provided for the lowering of the tariff, but failed to pass.

Next in tariff history comes a period of tariff reduction and reform embracing the years 1870–1883. The first Act of this period was that of 1870. This Act provided for reduction on some articles, but it also raised the duty on others. However, in 1872 a law was passed providing for a general reduction of 10 per cent on all goods; the free list was also increased.

Although the free traders were jubilant, the results proved a victory for the protectionists. With very little difficulty, in 1875, the protectionists caused the repeal of the reductions of 1872. Attempts were made in 1878 for reform but they were of no avail. On the whole, the high rates established during the Civil War remained up to 1883.

Then in 1883 came the first great revision since the Morrill bills of Civil War fame. The public demanded reforms, and Congress passed the Act of 1883. A commission was appointed to study the situation in 1882, and it reported in 1883. "Congress was controlled by protectionists, and the bill was a half-hearted attempt on the part of those wishing to maintain a system of high protection to make some concession to a public demand for a more moderate tariff system."[7] Some rates were lowered. This act remained until the beginning of the new era in 1890.

The next step was a sharp and unmistakable movement toward still higher protection. The McKinley Act of 1890 extended the protective system even beyond its great Civil War development. It was a simple statement of protection.

This Act remained in force until 1894, when its supporters received a crushing defeat at the polls. The Democrats came into power and passed the Act of 1894. This tariff sliced the duties on all imports, but essentially remained a protective measure. Wool was the only important item on the free list. The general range of duties was lowered, far below those of the McKinley Act, about to the level of those of 1883. As far as it went, this Act lowered duties, but in a faltering and feeble manner.

After the passage of this tariff nothing seemed so remote as a return to high and all-embracing protection; but the Act of 1894 remained in force only until 1897, when the Dingley tariff was passed.

The question of free silver was before the people, and the election of 1896 ousted the Silver Democrats. The first thing the incoming Republicans did was to pass the Dingley bill, which provided for protection in new directions and farther than ever before. The act was an outgrowth of an aggressive spirit of protection in the Republican ranks.

The next important step in the development of the tariff

[7] Taussig: "Tariff History of United States." Page 196.

was the creation of a Tariff Board of three members in 1909. These members were appointed by President Taft, and their duty was to study and investigate the condition of American industry and its relation to the tariff. Before they could accomplish much, the appropriation which supported them was cut off by Congress in 1912.

The Payne-Aldrich Tariff, 1909.—Meanwhile the Payne-Aldrich Tariff of 1909 was passed. It was a high protective tariff in some features. Ida Tarbell in her book, "The Tariff in Our Times," page 203, criticizes the bill most adversely. She believes it was the work of incompetent men guided by dishonest ones. She blames Taft for lack of leadership in not demanding real reforms from Congress. Many rates were lowered by the House; the Senate raised as many as possible.

The Underwood Tariff, 1913.—Cynical and sneering remarks from the public greeted the 1909 tariff. In 1912, due to the split in the Republican party, the Democrats came into power for the second time since the Civil War. There was a downward revision of the tariff, but from then until 1922 the tariff question was subordinated to the greater issues of the World War. Under Wilson was passed the Underwood Tariff, a tariff of reduction. During this period steps were taken to create a permanent board of advisers and experts on the tariff problem.

The Fordney-McCumber Tariff, 1922.—Immediately after the war came the election of 1920. The people defeated the Democrats on the foreign policy issue as expressed in the League of Nations. The Republicans gained control of all branches of the Government. They started work on a post-war tariff under the leadership of two "Old Guard" conservatives, Mr. McCumber of the Senate and Mr. Fordney of the House. They brought forth the bill which bears their name, and after lengthy discussions had it passed virtually as they proposed it. It was an extremely high protective tariff law. The elections of 1922 cut the Republican majority down, and gave place to a liberal bloc which worked to influence tariff cuts. The President, by the Fordney-McCumber Act, was authorized, if economic or other conditions warranted such action, to substitute for the foreign prices of imported goods the prices which the goods were expected to bring when sold in the American market, in determining the tariff which would be paid on such imports. It was thought that by this "elastic" provision the President could prevent dumping of foreign goods on the American market and the excessive increases in the prices of commodities.

The Hawley-Smoot Tariff, 1930.—The passage of the Hawley-Smoot Tariff on June 14, 1930, brought to a close one more chapter of the long and turbulent tariff episodes in United States history. In 1928 both Democratic and Republican parties advocated protection of industry. The Hawley-Smoot bill reached the highest protective level of any tariff law ever passed, with an average rate of about 20 per cent above that of the Fordney-McCumber bill. Various opinions concerning this bill have been expressed, many of which are in violent conflict. In the United States they ranged all the way from the prophecy of some that the tariff would ruin the export trade of the country to Senator Watson's assertion that after thirty days of the tariff's operation the country would find itself out of the prevailing business depression.

Recent Financial History.—The stories of the free silver agitation, the financial problems of the developing industrial nation, and the establishment and workings of the Federal Reserve System properly belong to a financial history of the United States. The story of the governmental control of the railroads also is hardly pertinent here. The railroads which had been placed under government control during the World War were returned to their owners March 1, 1920.

Industry During the World War.—The great power and capacity of industry showed forth during the World War. In addition to the governmental control of railroads, Federal commissions sought to bring all types of industries and all the resources of the nation into line for the purpose of winning the war. The deadly menace of the war was used as a justification for control and restrictions such as industry in the United States had never before known.

At the outbreak of the war in 1914, foreign holders of American stocks and bonds rushed to dispose of their securities. The stock exchange in New York closed in order to prevent a general fall in prices. Foreign exchange fluctuated. German trade was driven off the seas, and the Allies were engaged heart and soul in war preparations. The United States had lost the use of European boats. The emergency need for shipping facilities caused the Federal Government to act for relief. Cotton raisers were provided with an extended system of credits, to assuage the threatened loss due to loss of foreign markets and shipping facilities. A War Risk Bureau was also established by which the government undertook to compensate American merchants for the risks taken by engaging in foreign commerce during the war.

The demands of the Allies for food, supplies, and munitions soon became tremendous. American industrialists saw that their prosperity depended then upon the number of ships available to carry these items to Europe. The German submarine warfare, also, had cut the meagre supply of boats to an even smaller number. In September, 1916, Congress created the United States Shipping Board, the functions of which were widened when the United States entered the war. The Board had supervision of foreign commerce, control of vessels, and training of men for the merchant marine. In addition to these its largest duty was the executing of the shipbuilding program of the nation. In 1917 the Emergency Fleet Corporation was organized with a capital of $50,000,000, for the purpose of constructing merchant vessels. In four years, from 1916 to 1920, American tonnage increased about 14,000,000 tons. The Emergency Fleet Corporation built ships faster than the Germans could sink them. The disposal of these ships after the war became a national question. The Harding Administration planned to grant a subsidy to these vessels, but in March 1923 Congress abolished the Emergency Fleet Corporation, without acting on the proposed subsidy.

The Council of National Defense.—It is interesting to note the various agencies for war-time control in industry. One of the most important bodies of this type was the Council of National Defense. This council was authorized by Congress in 1916, and included the secretaries of war, navy, agriculture, interior, labor, and commerce. An advisory commission appointed by the President aided this body. Committees on such topics as coal production, shipping, and aircraft production also came into being.

Food Control During the World War.—The Food Control Act was passed in August 1917. The United States Food Administration and the United States Fuel Administration were organized to encourage the production of food and fuel and to supervise their distribution and consumption. The use of sugar, wheat, meat, butter, and other foods was restricted, and large quantities of food were diverted from home consumption to army use. The farmer was encouraged to increase his production of necessary foods; the era of the war garden in city and suburban grass plots came in. American people were asked to eat such grains as corn and rice, which they had never eaten in quantity before, in order to save the wheat for overseas consumption. Wheat has concentrated food value, and lends itself especially well to storage and shipment.

Another feature of the so-called Food Act authorized the President to purchase, in the name of the Government, certain grain and food products. Spring wheat was bought at the price of $2.26 per bushel at Chicago. This purchasing guaranteed a uniform price to the farmers. The newly established Food Administration Grain Corporation attended to this purchasing for the Government.

The War Industries Board.—The War Industries Board operated in cooperation with the Council of National Defense. It was created in July 1917 for the purpose of regulating production and purchase of war material. It also looked to the distribution of capital, material, and labor. It acted as a link between the industries and the government. The War Trade Board was a similar body whose field of supervision was foreign commerce. Money and credit were in the hands of the War Finance Corporation under the leadership of the secretary of the treasury. Through this body the government lent money to essential war industries. The corporation had the power to issue bonds and to deal in United States securities.

The War Labor Board.—The War Labor Board was organized in 1918. The war created a demand for labor, at the same time that it caused a shortage of the same. The effect of shortage of labor after 1914 can be seen in the increasing wages, the attractive offers made by many capitalists to laborers in competing plants, and in the number and character of strikes. The Labor Board was a court of appeals in disputes between employers and employees. It sought to prevent strikes, lockouts, and other methods of interfering with production. Mediation was the means by which the Board sought to accomplish its end.

Depression After the War.—The World War, as all great wars have done, interrupted the natural and orderly evolution of industry. The war period itself was one of untold prosperity for the United States. Europe needed goods, and America was the only country which could supply this demand in a large way. The exports to Europe in 1915 amounted to $1,971,434,000, while in 1917 they were $4,324,512,000. Along with the increase in demand, the prices of commodities advanced. Food, metals, manufactured articles, machinery—all rose to heights unknown.

This expansion of industry created an increased demand for labor. Many workers, also, left for service in army or navy, and thus left the labor supply decreased. Wages rose, and workingmen could drive hard bargains with employers in the boom market. Labor unions swelled in membership, and were prominent in almost all fields of labor activities. It was a period of strikes and threatened strikes. Labor, because of the demand on its services, usually found itself successful. The increased wages, however, made the worker a consumer also, and workers in industry joined in the era of rapid spending. The workers also were creditably represented in the lists of war loan subscribers.

The armistice in November, 1918, closed the World War.

Industrial conditions of the war period held over until the spring of 1920. Prices and wages continued to advance until this time, and business flourished. As a result, inflation took place. Farmers, gratified at the prosperity, purchased new lands, developed old fields, and invested in new buildings and improved machinery. Manufacturers also extended their plants, in many cases borrowing to do so. Prices for goods continued to rise; speculation was rampant.

Then, in order came a period of business collapse, stagnation, and, subsequently, recovery. The depression became acute in the last months of 1921 and continued until 1922. At the end of 1922 recovery was almost complete. The causes for the depression are deeply rooted in the effects of the war. Europe could not continue to buy in America. Her countries were burdened with staggering debts, and were unable even to meet the interest on their war debts. European money depreciated in value. Hence the purchasing power of European money diminished, and, as a result, those countries reduced their purchases. Foreign exports of the United States dropped more than $2,000,000,000 from 1920 to 1921. American manufacturers had produced for an anticipated demand that did not materialize. The result was an oversupply of goods in American markets and general business stagnation. High prices also caused domestic buyers to curtail the number of their normal purchases. The effect of this economy in purchasing, both foreign and domestic, was a sharp decline in prices.

The Farmers the First to Suffer.—The farmers were among the first to suffer. The falling off of export trade left great amounts of food stuffs on their hands, the prices of which were rapidly declining. Not only were the profits of the war period reinvested by the farmers, but further expansion by enterprising farmers in many cases was financed by mortgage loans. In the period of depression the farmers were driven to virtual destitution. The recovery of farm products prices was slow, and did not appear until 1923 to any noticeable extent.

Among the prices that continued high during the decline

period were rents. This was due to the housing shortage resulting from the interruption of building during the war.

Reduced operation first hit the automobile industries. Economy on the part of the people caused a marked drop in the demand for new automobiles. Unemployment then became a mark of this industry. The steel industry next felt the blight of depression and other industries were soon affected.

Industrial Revival, 1921–1929.—The revival in manufacturing was slow in coming. In 1921 there could be noted a resumption of activity, and by 1922 manufacturing was again increasing. The revival, though slow, was steady. By 1924 normalcy was again achieved. In December 1926 the increased profits led the United States Steel Company to declare a 40 per cent dividend on its common stock.

Labor felt the effects of the depression period. Unemployment was widespread. Surveys showed that, in practically all states and in all callings, employment in 1922 was less than in 1921. This unemployment, together with the general depression, caused a wage reduction. Strikes were no longer successful. Wage cuts were accepted, either by agreement or by defeat in strike.

The Era of Stock Speculation, 1929. The Wall Street Crash.—During the year 1929 the United States was the scene of widespread stock speculation. Steady rises in stock issues had characterized the years 1923–1928. In the spring and summer of 1929 the issues on the New York Stock Exchange and the Curb Market soared. Many stocks reached unheard of levels. Speculation in standard and questionable issues continued during the late summer. Bank presidents and office boys alike sought opportunities to increase their stores of money. And then, in late October and early November, the crash came. Within the space of a week stocks that had reached the four hundred and three hundred level were back around the one hundred level. Paper profits and marginal accounts were wiped out in what history will record as the stock market crash of 1929.

President Hoover Assumes Leadership for a Revival of Prosperity.—Soon after the slump, President Hoover called into a series of conferences the industrial and financial leaders of the country. There was undertaken a gigantic program of industrial and financial activity, and an attempt was made to strengthen the economic structure of the country.

A closing picture of American industry would show at the present time the rearing of an extensive revival and development of activity on a foundation secured by stability and wise financing.

Recent Immigration Legislation.—In concluding this brief summary of American industrial history a word is necessary on the recent immigration law and its effects on labor in the United States.

During the depression period of 1920-22, about five million workers lacked employment, but a real shortage of labor appeared when industry revived in 1923. Wages recovered, and, in many industries, almost reached the 1920 high point.

The 1923 shortage of labor was acute in the building trades. The suspension of building during the World War caused the workers of this industry to seek employment in other lines. Also, few recruits were received in this slack period. Then after the war, building went on at a greatly accelerated rate.

Cowdrick, on page 400, states:

"Certain classes of skilled factory labor were insufficiently manned. This was due in large measure to the partial breakdown of the apprenticeship system, which, in turn, was a result of the high wages paid for all kinds of labor during the war period. From 1916 to 1920 a youth had no difficulty in getting a job at unskilled or semiskilled labor, at which his earnings reached figures which destroyed whatever attraction there otherwise might have been in small earnings during a long apprenticeship in a skilled trade. Moreover, many boys had spent in military service the years in which they might have served apprenticeship."

Common labor was also exceedingly scarce. The pick and shovel men were few in comparison to the demand of industries for them.

America had, throughout its entire development, relied upon

Europe as its source of common labor. Due to various social attitudes and ideals, the American-born disliked the common labor of the pick and shovel. The recent immigrant found this type of work eagerly awaiting him upon his arrival.

After 1880 the stream of European immigration changed its source, and came more largely from southern and eastern Europe. Italians, Slavs, and Poles came in increasingly large numbers. The growth of immigration greatly alarmed thinking Americans. A glance at the figures will show a cause for this alarm. In 1900 the immigration was 448,572. By 1907 the number rose to 1,275,349. Again in 1914 it was 1,218,480. The war then stopped the general inrush, and during that period the yearly figures were about 100,000. After the war the flood commenced once again, and, in 1921, 305,228 aliens arrived. Indications of even greater numbers in future years and a realization that those who were coming were in a large part war cripples and destitutes brought forth governmental action.

The three per cent immigration law was passed in May 1921. This emergency measure provided that, in any one year, immigration from a country could not exceed three per cent of the natives of that country resident in the United States in 1910. In 1922 this law was extended for two years. About 350,000 were allowed in each year according to the regulations of this law. Northern European countries rarely filled their yearly quota, while southern Europeans rapidly filled theirs, and clamored for larger quotas.

In 1924 Congress passed a restrictive law in a permanent form. Instead of quotas based upon foreign born residents, a flat total of 150,000 was set for the quota countries, this total to be distributed among the quota countries in the same proportions as persons deriving their origin from each country respectively were found among the residents of the United States by the Census of 1920. This is called the principle of "national origins." This plan went into effect July 1, 1927. Up to that time the old three per cent law had been extended, save that the basis was changed to the census of 1890 instead of that of 1910.

Meanwhile American industry has had to work out a plan

to provide itself with labor. This has been done by wage incentives, mechanical improvements, efficiency devices, improved processes, and a wider and yet more intensive system of vocational education aided by Federal and state funds.

American industry of the future must be built upon a foundation made firm by science working in the fields of engineering, finance, and education.

Part Two

THE EDUCATIONAL EVOLUTION

Part Two traces the evolution of education from savage days to the present. Within the limits of a work of this nature only the briefest sort of review could be attempted. However, the great milestones in the path of progress have been enumerated and evaluated. It is hoped that their significance and influence in relation to vocational education will be noted.

For the story up to 1870, Charles A. Bennett's excellent source book, "A History of Manual and Industrial Education up to 1870," has been relied upon. Much use has been made of the remarkable treasure trove of ancient and modern documents reprinted in Bennett.

For the development after 1870, innumerable books, reports, and pamphlets have been surveyed. Of these perhaps Arthur Mays' book, "The Problem of Industrial Education," has proved most helpful.

THE EDUCATIONAL DEVELOPMENT

Early Times

Introduction.—Although no complete history of education is to be attempted here, it is necessary to correlate certain educational changes with the changes in industry already outlined. It is necessary also to study certain theories and practices which flourished during ancient, medieval, and modern times, as a basis for seeking the roots of modern vocational education. Certain leaders and their work must be studied in order to accomplish this purpose.

The Divisions of the Problem.—The problem divides itself into three major divisions, according to time. They are the ancient, the medieval, and the modern eras. It will be necessary to survey the work of the ancients very briefly, the work of the Greeks and the Romans, the work of the cathedral and monastery schools, and the guild schools. It will also be necessary to note the philosophy of the medieval universities and the influence of such men as Plato, Socrates, Aristotle, Christ, St. Benedict, Charlemagne, Alfred the Great, Peter Abelard, and the Jews.

In reviewing the work in education of the period of the Renaissance, a survey will be given of the cultural education and "Humanism" in Italy, the Court schools, Erasmus and the effects of the Renaissance in the North, the educational implications of the Reformation, and the Catholic reaction. There will also be mentioned the realistic movements of Rabelais, Montaigne, and Bacon. To complete the survey a sketch will be given of the work of Locke, Rousseau, Pestalozzi, Fellenberg, Niemeyer, Herbart, Froebel, Robert Owen, Horace Mann, and Henry Barnard.

Primitive Education.—In primitive society education is simple, direct, and definite. Primitive society presents a vivid picture of real vocational education, in which imitation is the method employed for the most part. Instruction has rarely occurred as such until the time when the primitive tribe has passed into the semi-civilized stage.

Primitive Religion.—In order to understand much of the education of the primitives it is necessary to be aware of dominant beliefs and cultural ideas. All savages are alike in one respect—they interpret their environment in much the same way the world over. This interpretation is known as animism. For every material existence or phenomenon the savage supposes a spiritual entity which controls or directs the material object. To all his belongings,—his weapons, his dog, his implements,—he attributes a double. This philosophy of animism explains his religious observances, his propitiations, and his hallowed rites.

Primitive Processes of Education.—Paul Monroe in his "A Text-Book in the History of Education" (page 4) lists two processes in the education of primitive man, as follows:

"The first is the training necessary to the satisfaction of the practical necessities of life. This training consists not alone in learning how to accomplish the object,—that is, to hunt, fish, use weapons, prepare skins, and secure shelter,—but, as well, how to do each of these things in that definite prescribed way which, through the experience of the clan or family,—as interpreted by the shaman, exorcist, medicine man, or whatever the functionary may be called,—has been found to avoid offending the doubles that preside over these material things.

"The second is the training in the elaborate procedures, or forms of worship, through which it is necessary that every member of the group shall go in his endeavor to placate the spirit world, or to cultivate its good will."

The first process is man's practical and vocational education, the second is his theoretical education. From the second has developed man's religion, philosophy, and science. It is the germ of man's spiritual and intellectual progress.

Primitive Division of Labor.—In primitive education the fundamental social institution, the family, is the sole educational institution. When life became more complex and the division of labor became a reality, the training process also became more clearly defined. The first division—that of the labor of man as against the labor of woman—is reflected in the different types of training for the boy and girl. At best even the training in the making of weapons, in war, in hunting, and in fishing was imitative. The training for the girl in the preparation of the food and shelter followed the same method.

Sample of Primitive Education.—In the "Handbook of American Indians" (U. S. Bureau of Ethnology, Vol. I, pp. 414–415) this paragraph is found:

"The aborigines of North America had their own system of education, through which the young were instructed in their coming labors and obligations, embracing not only the whole round of economic pursuits, hunting, fishing, handicraft, agriculture, and household work, but speech, fine art, customs, etiquette, social obligation, and tribal lore. By unconscious absorption and by constant inculcation, the boy and girl became the accomplished man and woman. Motives of pride and shame, the stimulus of flattery or disparagement, wrought constantly upon the child, male or female, who was the charge, not of the parents and grandparents alone, but of the whole tribe.— The Eskimo were most careful in teaching their girls and boys, setting them difficult problems in canoeing, sledding, and hunting, showing them how to solve them, and asking boys how they would meet a given emergency. Everywhere there was the closest association, for education, of parents with children, who learned the names and uses of things in nature."

A description of the construction of toy tools by the children follows:

"The Apache boy had for pedagogues his father and grandfather, who began early to teach him counting, to run on level ground, then up and down hill, to break branches from trees, to jump into cold water, and to race, the whole training tending to make him skilful, strong, and fearless. The girl was trained in part by her grandmother, the discipline beginning as soon as the child could control its movements, but never becoming regular or severe. It consisted in rising early, carrying water, helping about the home, cooking, and minding the children. At six the little girl took her lessons in basketry with Yucca leaves. Later on decorated baskets, saddle bags, bead work, and dress were her care."

These two examples are sufficient, in this very brief résumé, to indicate the general tone of primitive education.

Egypt.—In early Egypt vocational education was carried on in the temple colleges, where religious literature, ritual, and ceremonial were taught in the work of preparing priests for their various duties. The soldiers, another of Egypt's classes, received vocational education in private schools or in the army. The industrial class received its training for work largely through an apprenticeship system. As the Nile made Egypt, so architects and engineers were a necessary part of Egyptian society. They often received special training in the temple schools along with scribes, physicians, and singers. The caste system of Egypt limited the progress of education, and so allowed only few changes in the vocational education for surveying, engineering, architecture, and the decorative arts.

Jewish Education.—Prior to the Exodus from Egypt, the Hebrews were in the patriarchal stage of development and the family was the social unit of education. The father was the teacher of the boys; the mother shared the burden of instruction, especially of the girl. The aim of Jewish education was moral, religious, and vocational. After the Exodus, the Jews passed into the agricultural stage of development. The family organization was still the school unit, but training for the priests and for the scribes was now necessary, and the lay prophets needed special preparation for their work. These changes gave rise to the schools of the prophets. The vocational training consisted of training a boy for a trade and a girl for home-making.

When the Jews returned from the exile, they brought back with them the idea of a teaching institution outside the home, with a special class discharging the function of teacher. However, the new teaching class busied itself making a systematic study of the Torah, which had been reduced to writing, in institutions called synagogues.

On the Sabbath they expounded the Torah to the people at the synagogues, and they exercised judicial functions in cases arising under the Torah. Their decisions and commentaries upon the Torah, when collected later and reduced to writing, constituted the Talmud. They were trained for their work at a college, the House of Instruction, where they studied under reputable scribes. Interpretation of the Torah, and a study of Greek constituted the major studies of the higher courses. These scripture scholars became the teachers in the elementary schools for the people when these were later established.

Compulsory elementary schools were established in Jerusalem in 75 B.C. These grew out of voluntary schools established in the second century before Christ. These schools were moral and religious in aim, the vocational aim still being vested in the father. It is interesting to note the organization of the synagogue education. Only boys went to the synagogue school; the girls learned at home. There was compulsory attendance; rich and poor alike worked together, and no teacher had under him more than twenty-five pupils. Teacher qualifications included the following points: The teacher must be married, mature and able. The work was divided into two parts on the basis of age, one from six to ten years, the other from ten to fifteen. In the six to ten-year period, the content of instruction was reading, writing, and simple arithmetic. The reading text was the Pentateuch. The ten to fifteen-year period comprised training in the reading and the explanation of the Talmud.

Oral teaching and dictation with large appeal to the memory were the chief methods of instruction. The Talmud itself contains hints on pedagogy and notes the fact of individual differences and the use of mnemonic devices.

Corporal punishment was in vogue in dealing with refractory conduct and in dealing with backwardness in lessons. Rewards and prizes appealed to many of the pupils.

Greek Education.—In Homeric Greece education was practical and was the function of the family, the tribe, and the council. What little we know of this period is found in the Iliad and Odyssey of Homer.

In later Greek history there are two streams of influence to be studied—the Athenian and the Spartan systems of education.

Athens and Sparta Leading Greek City States.—Military excellence and public usefulness with the view to efficiency in time of war was the aim of the Spartan state. Athens, on the other hand, stressed individual excellence through physical and intellectual attainments.

In Sparta the state controlled the individual from birth to death. The content of Spartan education was predominantly physical training. The hardening of the boy and the development of the capacity to endure pain were primary aims. Running, jumping, wrestling, dancing and the like constituted a large part of the training. The moral element of Spartan education aimed at traits conducive to successful soldiering—respect for authority and self-dependence. Intellectual elements were few in Spartan education. Reading and writing were not usually taught. The chanting of selections from Homer to military tunes and drill in "laconic" speaking constituted about the only attempts at cultural development.

In the physical education of Athens, bodily grace and harmony were the aim. Graded exercises in the gymnasium were carefully selected and drilled. In moral education service to the city state was stressed and informal training was given the youth from sixteen to eighteen in civic duties through contact with the elders, and attendance at the assembly and at the juries. The intellectual content of Athenian education was provided by the music school. There training was given in music proper, in reading and writing, in poetry, and in dancing for aesthetic purposes.

In Sparta the method used was training rather than instruction. Participation in activities with other boys or mentors was the core of Spartan methodology. Tests were often given, but they were not bookish, and they tested habits and capacities rather than memories. Corporal punishment was inflicted for moral offences and mental inalertness. Emulation was extensively used, and fear of public disapproval was great.

In Athens the pedagogues had the right to inflict pain upon

the body of a pupil. Emulation was much appealed to, and serious infractions of moral conduct by a boy of the sixteen to eighteen-year period were dealt with by the Court of the Areopagus.

The education for women at Sparta had a strictly vocational basis. The home was the school. The aim was to develop mothers of healthy soldiers. Physical exercises, singing, and dancing for patriotic and religious purposes characterized the content of Spartan feminine education. In Athens education for the women was neglected. The household tasks were learned by the pick-up method, although the major duties were performed by slaves, who learned by that method.

Three Schools of Greek Philosophy. The Sophists.— With the advent of the age of Pericles (459–431 B.C.), three schools of philosophical thought influenced Athenian education.

One of these schools of thinking was the band of teachers called the Sophists. These came to Athens to meet the demands for a new education in keeping with the changed conditions of the new Athenian empire. The Sophists were colonial Greeks who had traveled widely and were familiar with the outside world. Their philosophy taught that man was the measure of all things. They were individualists. Their numbers included Protagoras and Prodicus, leaders in grammar word usage. The Sophists were opposed by the reactionary group because they were outsiders, because they accepted pay for teaching, and because their teaching aimed to prepare for personal advancement rather than for service to the state. The youth of Athens, however, flocked to the lectures of the "wise men."

The aim of the teaching of the Sophists was twofold. They stressed a moral and a vocational aim. They encouraged the students to work out their own adjustments to the mores of old Athens. They also endeavored to train public speakers at a time when oratory was the key to political advancement. Students paid fees to hear the Sophists. There were no school buildings. The lecture method prevailed, and grammar and rhetoric were taught as aids to forceful speaking.

The Reactionaries.—Mention has been made of those who opposed the Sophists. These have been termed the reactionaries. They sought to stem the tide of individualism which was rampant at this time in Athens. They attempted to return to the old ideal of city-state service. Pythagoras and Aristophanes were characteristic members of the group. The former stressed social co-operation, the latter ridiculed the Sophists in his comedies. Xenophon, who was also allied to this group, brought out the Spartan ideal of state service.

The Mediators.—The name, the "Mediators" is often given to the third group of thinkers in Athens of the time of Pericles. This group comprised the philosophers Socrates, Plato, and Aristotle. These men realized that conditions had changed, and that the old ideals and education no longer functioned or were even adequate to meet the new demands. They saw no solution in the individualistic philosophy of the Sophists. They believed that a new morality and education would have to be worked out to fulfil the new conditions of life.

Socrates.—Socrates left us his ideas on education through his pupil, Plato. That knowledge is virtue, and that there are concepts such as piety, temperance, courage, upon which all men agree, were characteristic tenets of the followers of Socrates.

The major aim of Socrates was to develop in the individual the power to think and to establish an agreement as to fundamental concepts.

His method was conversational. The first step was ironic. The pupil was brought from unconscious ignorance to conscious ignorance. The second step was constructive, and attempted to lead the pupil to truth.

Plato.—Plato was the pupil of Socrates. He was a native Athenian and the founder of a philosophical school. His educational contributions are found largely in the "Republic." Plato's aim in education was for virtue in the individual and justice in the State. Virtue in the individual is a result of the harmonious development of man's nature, appetites, passions, and intellect. Justice in the state is achieved when those best fitted to serve as artisans are serving as artisans, those best fitted to serve as soldiers are serving as soldiers, and so on. Plato approved of the prevailing Athenian content of elementary education, that is, music, gymnastics, and military training. He added arithmetic, geometry, astronomy, and music for the more advanced stages of education. The highest stage of education according to Plato was to be a study of philosophy.

Aristotle.—Aristotle, the third of the Mediators, was a native Athenian, the pupil of Plato and the teacher of Alexander the Great. His "Ethics" and "Politics" are replete with ideas on education. Aristotle's philosophy has been epitomized in the statement that virtue does not consist in knowledge, but in discharging the highest function for which an object is created. Man's highest function, he said, is to be rational in thought and in conduct. With this as a basis, he set up his threefold educational aim—to develop the three phases of man's nature: the body, the irrational soul, and the rational soul. The proper development thereof makes, he contended, for rational living, which is well-being and well-doing.

Aristotle believed in firm State control of education. He also believed in education for only the free male population. He stressed the home as an institution where effective early training habits must be begun. He divided the student class into three grades, from birth to seven, from seven to puberty, and from puberty to twenty-one years. He also includes, above this, the higher life of study and contemplation. The content in the lower stages included the formation of physical and moral habits, while in the higher stages gymnastics, music, and drawing were to train the irrational soul. For the period after puberty, training was to be gained through mathematics, logic, and the sciences.

Greek Education for the Cultured.—Although some of the Greek philosophers urged the study of practical subjects as a necessity for the cultured man, yet these were not to be used as a vulgar means to livelihood. Socrates warned his hearers against a study of philosophy with an ulterior purpose in view. In Athens refraining from work was simply an aristocratic virtue in contrast to the common industrial pursuit from which every educated person must be free if he would be respected. Plato identified labor and slavery. Naturally, with such a conception of labor and with such an attitude toward work no such development was possible in education for industrial efficiency as was found among Germanic nations of the latter middle ages, among whom the merchants and craftsmen became exponents of the culture of their time. The principle of utility could not be properly reconciled with that of harmonious development.

Professional Education in Greece.—However, preparation for the professions was not altogether lacking. Military training was intensely practical. The youths were taught the arts of war in the gymnasia and in the field much as our present day militia are trained. They were under vigorous discipline and were given intensive and extensive training in the art of war. A two years' service in the militia rounded off the soldier's training.

Medicine required a most direct preparation. Even Homer mentions how the sons of Aesculapius exercised their arts, one healing by surgery, the other curing by medicine. The cult of medicine, as practiced in the temples, was grossly superstitious, but must be regarded as the origin of the medical school. The famous schools of Hippocrates and Galen give witness to the degree of professional training given in the subject. Dentistry was a branch of surgery from the time of Herodotus. In Alexandria the schools of medicine made great progress, especially in anatomy.

Summary.—In summary, Greek education was undemocratic, and ignored the masses upon whom the state depended for its prosperity. Second, the end of education as a preparation for public life, while seemingly broad in scope and highly practical, was, upon closer observation, a narrow specialization. It was knowledge for the sake of knowledge. Third, the education was largely literary, ignoring the productive phases of the social life.

Education in Ancient Rome.—Roman education, to reflect Roman character and civilization, should have been above all practical. And so it was, until the adoption of Greek educational methods. Preparation for a vocation was the chief aim, accompanied by a practical scheme whereby family life, the battle field, the farm, and public and civic life co-operated. Theoretical education was subordinated to practical efficiency.

The Father in Roman Times.—In early times the father, as sole arbiter of his son's destiny, was under heavy moral obligation to provide him with practical and moral training. Individual teaching and apprenticeship was the method. Farm stations were established for those who were to become farmers; entrance into the army at an early age was the first step in a military career; observation of the methods of public life and of the patron's relations to his clients together with private or apprenticeship instruction formed the training for pleaders, jurists, and the public man in general; while, for the citizens engaged in arts and crafts, apprenticeship sufficed. Every occupation was provided for in the simplest and most direct way. The girl learned at her mother's side to spin, to weave, and to sew; the boy, in similar manner, learned the mysteries of planting, harvesting, swimming, riding, boxing, and the use of tools, both of peace and war. If the father was a priest, the son early assisted at the priestly duties. Education was moral, civic, and practical.

Roman Education Practical.—In "Education of Children at Rome" G. Clark (pp. 16–17) states that arithmetic was of particular importance. This was natural for a city of world-wide commerce and military supremacy. Arithmetic was the tool whereby the father and the merchant kept his important account book.

The Spread of Greek Ideals in Rome.—With increases in wealth, power, and domain there came a change in education and the adoption of Greek ideals. The practical Roman, like the conservative Cato, resisted the movement. The censors, in 92 B.C., decreed Greek education "Novelties," meriting disapprobation. In imperial times Roman youths in large numbers attended the philosophical and medical schools of Greece, and studied mathematics, astronomy, philosophy, and medicine at Alexandria. In Rome the school of Berytus was a veritable university of law. Schools of rhetoric were founded for the development of the orator, and these flourished for a while. The youth, after attending one of these schools, attached himself to some successful public man, whom he accompanied to the Forum, the Senate, or the Assembly, by way of apprenticeship. Soon, however, the rhetorical schools degenerated, got out of touch with real life, and became purely literary. The moralizing influence of the home was gone, too, and the decline and fall of the Empire was not far distant.

There are but few records left of Roman education, and in these little mention is made of professional education, except in the case of the special schools which trained for public life. The inference is that the architect and engineer mastered the technique of his intricate calling in some form of apprenticeship.

Roman Texts.—It is interesting to note some of the technical texts of the time. Cato, Columella and Varro wrote on agriculture, Seneca and Pliny on natural philosophy, Mela and Solinus on geography, and Vitruvius on architecture and engineering. In construction engineering, the apprenticeship system prevailed.

The Middle Ages

Christian Education.—With the fall of Rome and the inroads of the barbarians of the fifth century there is little to be said of education.

Christian education was moral, religious, and vocational. The first great advance after the fifth century was made under the guidance of Charlemagne and Alcuin. The rise of monastic, parish, cathedral, or episcopal schools, chantry or song schools, and "A B C" schools need be but mentioned. They were vocational in so far as they prepared for the priesthood.

Two Types of Early Christian Schools.—Two types of schools arose early; the catechumenal for instruction of converts, and the catechetical for the development of defenders of the church. For a study of vocational origins, the latter is the more important. Christianity was not long recognized before it was challenged by representatives of the pagan philosophy. Strong defenders were needed not only to propagate the faith but to preserve it. The catechetical schools sought to produce such defenders. They were indeed vocational schools. Here Christian teachers and leaders were instructed in theological doctrines and pagan philosophy, literature, and science that they might defend the faith, combat heresies, and formulate Christian doctrine. The direct method of the philosophical schools was followed; the aim and method were fundamentally practical.

The episcopal or cathedral schools were also practical and vocational, serving for the training of the clergy. Instruction in doctrine, ritual, the seven liberal arts, singing, and the duties of public religious service constituted the core of the curriculum.

The Feudal Age.—With the coming of the feudal age, each family became self-sufficient upon its manor estate. There was little or no manufacturing or trading, agriculture being the most secure way of subsistence. Life itself was insecure. The Norsemen or the robbers were prominent and effective characters of the period. There could be little education in such an age.

The Contributions of the Early Monks.—Labor among the early Christian monks occupied a significant position. Based upon the teachings of the rabbis of pre-Christian days and the example of the carpenter of Nazareth, it is not altogether strange that the educational practices of the early monks laid great stress upon manual labor.

In the "Select Historical Documents of the Middle Ages," by Ernest H. Henderson, is found the following extract from the Rule of St. Benedict (Quoted from Bennett's "History of Manual and Industrial Education up to 1870," pages 24–25):

"48. Concerning the daily manual labor. Idleness is the enemy of the soul. And, therefore, at fixed times, the brothers ought to be occupied in manual labor; and again, at fixed times, in sacred reading. Therefore we believe that, according to this disposition, both seasons ought to be arranged, so that from Easter to the Calends of October, going out early, from the first until the fourth hour, they shall do what labor may be necessary. Moreover, from the fourth hour until about the sixth, they shall be free for reading. After the meal of the sixth hour, moreover, rising from the table, they shall rest in their beds with all silence.—If the exigency or poverty of the place demands that they occupy themselves in picking fruits, they shall not be dismayed; for they are truly monks if they labor by their hands; as did our fathers and the apostles."

"57. Concerning the artificers of the monastery. Artificers, if there are any in the monastery, shall practice with all humility their special arts, if the abbot permit it.——"

Each monastery was a school of labor, and at the same time a school of charity. St. Benedict (480–543) founded the order of Benedictines in 538. He firmly believed in the Benedictine theory of labor, but he added the ideal of beauty to that of utility. He also saw the great opportunity of spreading the teachings of Christ and the church through books. He is credited with having originated the "scriptorium," a workshop devoted entirely to the production of books.

The influence of the monks spread rapidly northward. Soon Germany was dotted with monasteries which became centers of civilization through the teaching of husbandry and art and literature. The monks built their own monasteries, cleared their own farms, and did all the work necessary to the production of a real center of civilization.

Though its primary function was not educational, monasticism was in fact a system of education in which the hands played an important part. A monastery was a school with prescribed hours for work, reading, worship, and rest. Its discipline and regulation were in the hands of the abbot. Men who sought an opportunity for reflection and study flocked to the halls. Young and old felt its attracting force. Charles A. Bennett, in "History of Manual and Industrial Education Up to 1870" (pp. 20–21), states:

"Thus the monasteries came to be the schools for teaching, the place of professional training, the only universities of research, the only publishers of books, and the only libraries for the preservation of learning; they produced the only scholars; they were the sole educational institutions of this period."

However, the larger part of the monastic institutions offered the study of religious writings only. Some few did offer the "Seven Liberal Arts"—grammar, rhetoric, dialectics, arithmetic, geometry, music and astronomy. Soon, throughout Europe, the monastery had become the center for the instruction of boys.

The question and answer method was in general use and dictation by the teacher was a common practice. Memorization played a large part in the instruction of the boys. The customary age of admission to the monastery school was ten years; to the order, eighteen years. Many schools took boys who did not intend to take monastic vows.

Convents for girls corresponding to monasteries for boys grew up also. They taught reading, writing, and the elements of learning. The weaving of church hangings and the embroidering of altar cloths and vestments constituted a vocational side of their training.

The Educational Revivals from 500–1100 A.D.—Two great educational revivals mark the period between 500–1100. These were the revival under Charlemagne and that under Alfred the Great.

Charlemagne.—Charlemagne (742–814) aimed to reach a cultural unity among the people of his empire in order to reinforce the religious and political unity which had been achieved. He sought to utilize the monks, the monasteries, and the cathedral schools as a means to his end. He established a circuit palace school under the supervision of Alcuin of York for the education of the nobles. Next he issued capitularies to abbots and bishops calling for a better-educated clergy. Finally he sent inspectors (*missi dominici*) to the schools to see that the capitularies were carried out.

The results of Charlemagne's revival can be briefly summarized as (1) a renewed interest in continental Europe in classical education; (2) a higher standard of education among both secular and regular clergy; (3) the establishment by bishops of parish schools for elementary schools where they did not exist already.

Alfred the Great.—Alfred the Great (871–901) desired to raise the standard of education to the position it had held in the time of Alcuin, after it had fallen into decay due to internal wars and the invasions of the Danes. Alfred sent to the Continent for monks to assist him in this work. Like Charlemagne, he established a palace school for the training of the nobles. In this Saxon was taught as well as Latin. Monasteries were stimulated to better their schools. Alfred, himself, set about translating various religious and historical works into the vernacular. During this period the Anglo-Saxon Chronicle was begun. The results of Alfred's mark include an increase in the number of monastery schools, the foundation of many parish schools, and the appearance of English prose writing.

The School for the Knight, a Feudal Institution.—Another phase of education of a vocational character in medieval times was the practical training of the knight. His education was divided into two parts—that of the page from the seventh to the fourteenth year, and of the squire from the fourteenth to the twenty-first year. Every important land owner sent his son to the castle of the neighboring lord, or king, where obedience and service were instilled as the first elements of knightly virtue. Of special service was the page to the lady of the castle under whose care he was. Later, as squire, the youth accompanied his lord in the hunt or in battle. Tilting, jousting, and the other martial exercises were much practiced; tournaments were common, and where many squires came together it might be said that a school was formed.

Learning outside the monasteries or the feudal castle was relatively unimportant during the early decades of the feudal era. Participation in skilled labor was the principal means of education outside the abovementioned agencies, although it was not recognized as such by the schools. As the crafts became more differentiated and specialized in their development, there grew up an increasingly large amount of technical knowledge and manual skill to be gained before successful entrance could be gained into the craft. The apprenticeship system developed then into a complex educational institution for the middle-class youth.

Apprenticeship began in the home when the father taught his son; when he added another man's son he obligated himself to treat him as his own. The apprenticeship period usually covered seven years. The master was supposed to give his charges moral, religious, and civic instruction. He was to teach him all that was necessary for the craft, including receipts, rules, skills, and such applications of science, mathematics, and art as might be pertinent. The imitative method was usually employed. In certain crafts the guild forced the master to include reading and writing among the subjects taught.

The Universities.—The rise of universities with specialized professional faculties was the first protest against monastic education. (Their origin as guilds of students seeking secular knowledge is a probable outgrowth of the Crusades.) Their early organization and development is a familiar story. In passing, it may be noted that Salerno, a health resort amid mineral springs, specialized in medicine as early as the ninth century; Ravenna and Bologna specialized in law; Padua, in law, medicine, and philosophy; Seville and Salamanca and Paris, in theology. The twelfth and following centuries saw the establishment of the above-named and many other similar institutions. It was customary for a youth after completing the university training to seek employment for a practice period with a professional who had an established practice.

The Saracens.—It is necessary to mention briefly the great achievements of the Saracens in mathematics, chemistry, physics, astronomy, medicine, and surgery because of the influence they exerted upon European professional education. This was another indirect effect of the Crusades. The Saracenic development of algebra, arithmetic, trigonometry, and the use of the Hindoo method of numerical notation were of especial importance to medieval educational institutions. In the physicochemical field they made great studies. They discovered how to manufacture nitric and sulphuric acids and alcohol. They propounded laws of gases, they measured the specific gravity of many elements, they formulated laws of light and vision, and they developed the laws of the pendulum. In medicine and surgery they studied the properties of drugs and the technique of operations. They also compiled treatises upon their experiments and discoveries. The Saracenic medical schools established in Spain were frequented by students from all parts of Europe.

The Guild and Burgh Schools.—All through the latter part of the Middle Ages there can be noted a movement toward secularization of education. This is especially reflected in the developments known as the guild and burgh schools. The whole educational movement runs parallel to the rise and growth of trade and industry. It is also a result of the greater wealth of the craft and merchant guilds and hence their greater independence. The towns at this time were buying or wresting their charters from the feudal lords.

The evolution of this movement was slow. Its beginnings can be found in the guild school, which grew out of the feudal custom of maintaining one or more chaplains whose duty was to instruct the manor children in the principles of religion. Later, to his clerical duties were added the duties of instructor to the children of guild members. The elementary guild schools taught the "three R's," singing, and geography. The higher guild schools taught grammar, literature, and logic, with an aim toward university preparation. The guild schools of Venice, Florence, and Padua specialized in art, and in the seventeenth century became academies of art.

When the merchant guild became identified with the burgh government, the guild schools became burgh schools. The Merchant Tailor's School at London today is one example of a lasting school of this type.

Rich and poor went to the burgh schools, and occasionally very bright boys were taken from among the serfs and trained to be clerks, lawyers, or even bishops.

In Germany and northern Europe generally, the founding of the burgh schools was more direct. The town simply took over the parish school by patronage. It furnished the buildings, and paid the teachers' salaries. The movement met with the opposition of the church, as can be expected. The Hanseatic League was a prime mover in the establishment of schools to meet the needs of the new economic and social conditions.

The Renaissance.—The education of the Renaissance, though often characterized by a narrow intellectual discipline, bore the seeds of modern science. Its influence deeply affected thought and found material expression in the various inventions and discoveries which led ultimately to the Industrial Revolution. The education of the Renaissance, and especially that phase of it from which modern science has evolved, relied upon experimental method and observation of nature. Attendance at the humanistic schools of the Renaissance was a necessary preliminary for entrance to the learned professions, and their training was not for a vocation but for "polite living." Many were comparable to the finishing school of today.

The Reformation.—The Reformation was a natural outgrowth of the Renaissance. Its tendency was to swing the educational aim towards theology and morality. Reason and the importance of the individual as an instrument in his personal salvation would appear at first sight to be possible cornerstones upon which to rear a real educational structure. The result, however, was formalism, with the humanistic content worked over for theological ends. Religion dominated education until well into the nineteenth century.

Luther on Education.—Luther demanded more than a merely religious training, however. Education was to be compulsory for both girls and boys. Two hours' schooling a day was to be the limit for older children, the rest of the time to be devoted to the learning of a trade at home. Throughout his writings, Luther stressed the necessity of coordinating industry and the school.

The Roman Church and Education During the Reformation.—During the Reformation the elementary school was revived. National systems were established in Wurtemberg and Saxony in the sixteenth century and in Scotland in the seventeenth century. In Weimar in 1619 compulsory education was adopted. The Jesuit Schools, the Port Royal Schools, and the Schools of the Christian Brothers were dominated by religion. The Jesuits in 1599 began the training of teachers.

The first institution for the training of elementary teachers was opened in 1685 by the Christian Brothers.

The Attempt to Revive the Classics.—Though humanism had become formalism, there were still reformers who sought to revamp the classics so that they might lend themselves to the preparation for life. Erasmus, Rabelais, and Milton emphasized ideas and knowledge and attempted to use the classics to teach agriculture, natural philosophy, geography, and medicine. Montaigne largely rejected the classics. In Germany, under the old humanism, the Ritterakadamion prospered. The common schools of the people were nearly extinct by the last half of the seventeenth century. Under the philosophy of Hobbes and Machiavelli, the masses were born to be governed.

The Sense-Realism Movement.—The reaction against humanism brought with it the whole movement of sense-realism. Summed up in the philosophy of Francis Bacon, it stated that all things come through experience, investigation, or experiment, and that knowledge is for the use and relief of man's estate. Under these principles, Ratko in Sweden and Germany undertook to teach the arts and sciences. The encyclopedic idea was prevalent. Comenius held this idea, and attempted to teach something of breadbaking, fishing, weaving, tailoring, cooking, brewing, etc. in the "Orbis Pictus."

The Realschulen.[1]—As a direct result of this sense-realism movement arose the Realschulen in Germany influenced by Frederick the Great. Francke, in 1694, founded his famous institution at Halle to teach the arts and the crafts. The direction of thought was to prepare boys for the practical affairs of life other than the professions. A teacher's seminary was included at the institution of Halle. The Rostock Academy, founded in 1619, was actually the first Realschule. This was just after Kepler's famous discoveries. Principles of handicraft and manufacturing were taught at Halle, when Semler in 1706 opened a mathematical Realschule there. Hecker in his "Economic-Mathematic" School at Berlin in 1747 pledged to his pupils a preparation which would facilitate their entry into any trade.

[1] German Higher Schools, J. E. Russell, pp. 54–55.

Many of the burgh schools developed into Realschulen, and in England a number of academies arose. The universities resisted the movement, and education largely remained humanistic until after the time of Jean Jacques Rousseau.

The End of the Static Education.—The development of a sound theory of sense-realism follows in the history of educational evolution. Voltaire demolished the old static education, and Rousseau constructed the basis for the new. Emile, in his natural education, learned a trade for economic reasons. Basedow, a follower of Rousseau, included the following subjects in his Philanthropinum at Dessau: man, animals, trees, plants, garden and farm implements, mineral and chemical substances, mathematical instruments for measuring and weighing, the air pump, siphon, etc., the use of various trade tools, history and commerce, not to prepare for entrance into the trades but to train the rich for leadership in life, and the poor to be teachers. ("Educational Reformers," Quick, page 280.)

Modern Times

The First Industrial School. The Work of Pestalozzi.—Education for actual livelihood was the conception of Pestalozzi. On a farm at Neuhof from 1775 to 1780 he conducted an industrial school for the poor. This was the first real industrial school. He worked with twenty children, teaching them agriculture, weaving, and spinning, and the elements of learning at the same time. To Pestalozzi, a school that was not an apprenticeship to a livelihood was an absurdity. The school may have been an administrative failure, but it was certainly a landmark in the history of practical education.

Pestalozzi Influenced by Rousseau.—Heinrich Pestalozzi (1746–1827) was an intelligent reader of Rousseau's "Emile." At first he allied himself with the ardent revolutionists. He abandoned in turn his preparation for the ministry, for the law, and for public service, and finally entered upon an agricultural life. He purposed to improve a waste tract of land through new methods of cultivation and to live a life in accord with the prevalent naturalistic ideals. He failed to achieve practical success, but the failure gave him an opportunity for trying an experiment even nearer his heart's desire—the founding of a philanthropic institute for destitute children at Neuhof.

"The development of the factory system of labor had already begun to accentuate the economic division of the people and to produce a poverty-stricken class, whose children were much more neglected than those of the peasantry, and of whom no care was taken save by the poorhouses or charitable institutions that but increased the moral and industrial evils." (History of Education, Paul Monroe, page 602.)

Pestalozzi at Neuhof.—From 1775 to 1780 Pestalozzi conducted at Neuhof what was probably the first industrial school for the poor. The children were engaged in raising special farm products, in spinning, and in weaving. While so engaged, they also spent some time in reading and committing passages to memory, and especially in arithmetical exercises. There was no real connection established between the occupations and the intellectual activities, but Pestalozzi demonstrated that the two could go on together. Pestalozzi, the reformer, failed to fill the rôles of manager, farmer, manufacturer, merchant, and schoolmaster. The school population was practically the refuse of society. The parents were people who lacked the appreciation necessary for the survival of the enterprise. It was abandoned.

During the next eighteen years Pestalozzi participated in the revolutionary movement, and devoted himself chiefly to literary work.

In 1798 a complete change took place in the man. He decided to give up his theorizing and turn schoolmaster. His career as a schoolmaster has had more influence than any other one factor upon the educational improvement of the nineteenth century. One chief reason for this was that his ideas were the results of experimentation.

Pestalozzi at Stanz.—Pestalozzi accepted the charge of those children in one of the districts of Switzerland who had been made orphans through the massacre of the people by the French soldiery. The germs of new educational practices were worked out with these orphans at Stanz. Here again, as at Neuhof, his fundamental purpose was to combine educational activities with handwork. But he saw that if an approach differing from that of the ordinary schoolroom were made, much of the experience that was most valuable for mental development came directly from those activities in which the children were immediately interested. The keynote was simply that the essential to reform is a new method and new spirit in all educational work. After one year, however, the ravages of war terminated the experiment.

Pestalozzi at Burgdorf.—Next year, 1799, Pestalozzi was an assistant teacher in the village school of Burgdorf. Here he developed the object lesson. His plan to "psychologize education" met with little approval from the villagers.

In 1801 there appeared his most systematic work, "How Gertrude Teaches her Children." It was an attempt to answer the question, "What knowledge and what practical abilities are necessary for the child, and how can they be furnished to the child or obtained by him?" While working out his thoughts, he was aided by an endowed private school where for four years he experimented with pupils and teachers. When the government withdrew the endowment, Pestalozzi withdrew to Yverdun, where he carried on his longest and last attempt at experimentation.

Pestalozzi at Yverdun.—For twenty years Pestalozzi labored at Yverdun. His work was directed toward the training of teachers and direct experimentation in reforming educational practices. The work he began at Burgdorf was continued. Visitors were welcomed from almost every civilized land, and students were trained for various European countries. Text books were compiled, explanatory and controversial treatises were printed, and results were published. The old enthusiast, however, failed as a practical administrator, and his staff fell into internal dissension.

Pestalozzi's Contributions.—Pestalozzi contributed much to education. He was convinced, as were most great educators from the time of the Renaissance, that education was the key to social reform. In the period of Pestalozzi when atheism, anarchism, socialism, naturalism, pure individualism, and communism were advocated, Pestalozzi was heard above all others advocating education, a new education.

Pestalozzi presented a new meaning of education, a new aim, and suggested new means and methods. He demanded universal education for the masses. He believed in education as the organic development of the individual, mentally, morally, and physically.

Summary.—In summation—religious and humanistic education prevailed in this period. The major points of note are the work of the Benedictines, the training of the knight, the professional training of the universities, the specialized guild and burgh schools, the demands of Luther, the scientific and realistic movements starting with Bacon, the activities of the humanistic reformers, the rise of the Realschulen, the views of Voltaire and Rousseau, and the institutions of Basedow and Pestalozzi.

Early Industrial Schools in America.—At this point, there can be noted that schools for teaching industrial occupations along with the usual academic subjects of the time were taking root in the United States. In 1745 the Moravian Brethren established a settlement at Bethlehem, Pennsylvania. These followers of John Huss maintained a community similar in many respects to ancient Sparta. The children remained with their parents until they were about twelve years of age, when they were sent to a public school. This school is described in "The Knickerbocker or New York Monthly Magazine" published in 1849. In volume 34, page 24, it is noted that the boys were taught "reading, writing, arithmetic, the learned languages and other parts of literature according to their abilities, and the business they are designed for." Here likewise they were instructed in the elements of religion. At twelve the boys were sent to live at a house where all the single men lived, and they became subject to rigid discipline. They ate, slept, and worked together. The elder men were, for the most part, skilled artisans, and worked at their trades; the boys were instructed by these men in the trades and arts which they intended to pursue. A garden behind the house served for a laboratory for embryonic gardeners. In the house of single women, for these too lived together, the girls learned "spinning, knitting, weaving, needlework, embroidery, tambour, and other female arts" (Knickerbocker Magazine).

Agriculture an Early Subject for Training.—The Methodists established Cokesbury College in 1787 at Abingdon, Maryland. Here there was carried on a type of gardening and carpentry work that served as a model for much of the work later found in the manual training movement forty years later. In Bernard Steiner's "History of Education in Maryland" (Circular of Information #2, 1894, U. S. Bureau of Education) this college is described as a place where:

"A person skilled in gardening was appointed to oversee the students in their recreations, where each was at liberty to indulge his own particular taste from a tulip to a cabbage. There was also a place for working in wood with all proper instruments and materials, and a skillful person appointed to direct the students at this recreation." [The prospectus of the institution for 1785 explains this work]— "We prohibit play in the strongest of terms. . . . The employments, therefore, which we have chosen for the recreation of the students are such as are of greatest public utility—agriculture and architecture."

The Columbian Magazine for April 1787 contained an anonymous article setting forth a plan for establishing schools in a new country where the inhabitants are thinly settled and where the children are to be educated with special reference to rural life. The plea of the article was to prepare by education superior farmers and farmers' wives. The plan was to employ the children so that their labor would pay for their board and for a superior type of schooling. Each school was to have a tract of meadow, tillage, and woodland on which could be erected a house, barn, workshop, and school. The boys were to be taught growing, brewing, managing cattle and bees, etc. In the winter their time was to be employed in making "utensils of husbandry that would be requisite for the ensuing season." The girls were to be taught to sew, knit, spin, cook, keep house, and milk cows and make butter and cheese. Academic subjects were to be included also, especially geography, history, English literature, bookkeeping, geometry, and surveying.

The influence of this article in the educational thought of the time must have been considerable, for ten years later (1797) Dr. John de la Howe of South Carolina left a will which provided for the endowment of a school as proposed in the Columbian Magazine. The de la Howe State School of South Carolina is probably the oldest agricultural school in the United States.

The Work of Fellenberg.—Practical school organization and administration found a new and vital force in the contribution of Fellenberg. Dr. Henry Barnard in 1854 said that Fellenberg, in applying the principles of Pestalozzi, had attracted more attention at Hofwyl and exerted a wider influence than any one institution in Europe or America at that time. It was a veritable holyland to which pilgrims from the educational world flocked. Numerous are the reports of statesmen and educators on the results of their pilgrimage.

Manual labor was the distinguishing characteristic of the Fellenberg scheme, and hence the impetus of the venture is easily noted in the agricultural school, the industrial reform school, and the manual labor school.

Fellenberg at Hofwyl.—The philosophy of Fellenberg disclosed an interesting outlook on life. Education, he believed, needed to be reformed. Divine Wisdom has shaped the order of things so that some men are born to rule, others to obey. Hence education must be such as will fit each for his appointed place. He classified society into three castes, the higher, the middling, and the poor. The greatest defects in education were in that for the middling and the poor. While he believed in doing all he could for the advancement of the lower classes he firmly believed that the only rational course was to prepare them for the station in which Providence had placed them. All classes should be brought up side by side, so that the lower should respect and love the higher, and the higher appreciate and have sympathy for the lower. In 1799 Fellenberg purchased an estate of some 600 acres at Munchenbuchsee, which he named Hofwyl. Here began the great work which he had planned. He devoted his time to the improvement of agriculture, aided by private tutors. The growth of Fellenberg's Academy was slow and it was not until 1807 that the first school building was erected. The school was primarily for the children of the rich and it was conducted upon the social theories of Fellenberg. Robert Owen,[2] who was a student there, states that there was developed a very comprehensive system of student government.

By 1819 there were more than 100 students and from twenty-five to thirty instructors. Students came from England and Russia as the fame of the school spread. Board and tuition cost between $500 and $1,500. (Griscom, John: "A Year in Europe.") The course of instruction in the academy included the Greek, Latin, German, and French languages and literature, history (civil and sacred), geography, mathematics, natural and mental philosophy, chemistry, gymnastics, natural history, and religious instruction. Classes numbered from ten to fifteen. Private instruction was combined with class instruction.

Fellenberg's Farm School.—Fellenberg, with the aid of one Wehrli (or Vehrli), laid the foundation for the farm school. In this division the boys were clothed as farmers and fed upon a vegetable diet. Practical work in the field was interrelated with academic and devotional exercises. The whole work of the farm was carried on by the pupils, and a household council saw to the rotating of the various domestic jobs. Wehrli made a practice of forming as many connections as possible between school studies and manual work. While on the field the pupils received some of the most vital lessons in geography, history, natural science, and geometry, as well as in religion and morals. Fellenberg, himself, said: "Instruction should be followed by action as closely as the lightning by the thunder, and the life should be in complete harmony with the studies." (Quoted in "Sketches of Hofwyl," by William C. Woodbridge, Boston, 1831.)

[2] Threading My Way, Robert Dale Owen, 1874, given in Bennett's Source Material, page 154.

Fellenberg and the Mechanic Arts.—Although Fellenberg stressed the practice of agriculture as the best means of cultivating the happiness of men, he listed the mechanic arts second. There were employed at Hofwyl mechanics representative of the several different trades and each was provided with a shop wherein he could work. There was a blacksmith, a wheelwright, a carpenter, a cabinet maker, a turner, a brass worker, a shoemaker, a harnessmaker, a tailor, a lithographer, and a bookbinder. A young man could readily select and learn a trade.

Fellenberg was interested in the manufacture of new and improved farm implements. The surplus of those manufactured by the pupils were sold to neighboring farmers. The income therefrom was to assist in defraying the cost of instruction. In order to defray tuition charges, the poorer boys who paid no tuition were required to remain in the school until they were twenty-one years of age.

Due to the fact that at twenty-one a boy from this school was an intelligent practical farmer, skilled at a trade and possessing a general education, the demand for graduates far outnumbered the supply.

The handwork taught in the academy was so presented to the learner that he could learn only by imitation. The various jobs had not been subjected to scientific analysis, and therefore the methods bore little resemblance to the methods of our more advanced trade schools of the present era. All the evidence available seems to indicate that imitation was the prime method of instruction in the academy.

Fellenberg's School for Girls.—Under the direction of Fellenberg's oldest daughter a branch of the farm and trade school was organized for girls. The girls were taught to read, speak, and write their own language and to make such mental calculations as were considered necessary for them. The manual work taught included form and linear drawing especially applied to cutting out and making articles of clothing, spinning, knitting, sewing, cooking, washing, and care of the home. They also engaged in light farming tasks.

The Social Theories of Fellenberg Reflected in His Educational Activities.—In applying his three-class system of society, Fellenberg maintained his farm and trade school for the poor, the academy for the rich and an applied science school or practical institution for the middle class. In a special building he lodged the middle-class youth where they could live in a more simple manner than the boys in the academy and yet partake of its lessons. According to their proposed life work, they placed themselves in the shops, on the farm, or in the business office of the institution.

As Fellenberg gave prominence to agriculture, the greatest number of pupils were enrolled in the farm division. They received instruction in the art and science of farming. From what little evidence exists as to just how they carried this on, it is believed that efforts were made to treat from a scientific standpoint such subjects as weeds and their destruction, the breeding and care of cattle, the use of manures, and other methods of working the soil. (From "A Year in Europe" 1818–1819, by John Griscom.) In other words, experimental work in the field of agriculture was developed.

Fellenberg and Teacher Training.—Such an institution led to a demand for facilities for properly training teachers to carry on so wide and varied a program. In Fellenberg's institution there grew up a normal school to meet the demand. In the first year forty-two teachers of the Canton of Berne met and received gratuitous instruction in the art of teaching. The following year greater numbers appeared, so that the rulers of Berne, autocratically, forbade their teachers to attend these instructions under penalty of losing their positions.

Fellenberg then limited the field of the normal school so as to include only those employed at the same time as laborers at the school farm under Wehrli. Great stress was placed upon the principle that teachers must be familiar with the practical labor which is encountered by the pupil. Teachers were, therefore, instructed in the subject matter, methods, and general management of the school.

The Spread of Fellenberg's Ideas.—To impress upon the educational world the practicability of the Hofwyl venture, Fellenberg set to work to establish smaller school colonies. This was necessary despite the many failures of similar experiments. Isolation and the need for cooperation were set forth as guides in the founding of such establishments. Fellenberg also stressed the point that the institution must be small enough for each individual to see the results of his effort. In 1827 he established the school at Meykirch as an example of this plan.

The period immediately following the experiments of Pestalozzi and Fellenberg was one in which the beginnings of hand work in the schools developed rapidly. Public interest was centered in the extension of education to the working classes of society. The period from 1770 to 1870 was one pregnant with social and industrial change. The American and French revolutions, the application of steam power to industry, the development of machinery—all these and many other events filled this changing period. Adam Smith published his "Wealth of Nations" in which he pointed out the social value of the laborer and his labor. Education had to keep pace with social theory, and great were the numbers who sought education. In England, the Lancastrian system was developed whereby one teacher could manage great numbers of pupils. Social economics demanded education for the masses. Rousseau's ideas, modified and worked out by Fellenberg and Pestalozzi, were coming into practical effect.

Fellenberg's Followers.—Among the followers of these great educators of the early days was Heinrich Heusinger (1766–1837), professor of philosophy and pedagogy at Jena. His pamphlet, "Uber die Benutzung des bei Kindern zuthätigen Triebes beschäftigt zu sein" (quoted in Bennett), maintained that the child's impulse and desire for activity should be the key to education, and he made manual work the central point of his system. Froebel built up his system a few years later upon the principles set forth in Heusinger's pamphlet.

In the novel, "Die Familie Wertheim," published in 1798, Heusinger sets forth practical applications of his ideals. He believes that the work should correspond to the powers of the child; that the work should not be unhealthful, that the work should be executed while the child stands, that the work should be not merely the foundation for artisan's work, but for general education; that the materials used should be various in character; and that a sense of beauty should be developed through self-expression.

Niemeyer at Halle.—In 1799 August Niemeyer (1754–1828) of the University of Halle published a book on the principles of education, setting forth the idea of the harmonious development of the faculties of man. This, he recommended, was to be accomplished by the use of the manual arts. Carpentry, he said, is very suitable for the purpose because of the complexity of the tools and because it does not unduly strain the young. Gardening was of prime value. The young gardener gains health in the out-of-doors, works in communion with Nature and learns her laws first hand, exercises patience and witnesses a creation of his own, and through his manifold and varied experiences gains complete development. Above all, he says, the young should be familiar with the ordinary tools of the household for which there is a constant use. To keep these things out of the child's reach is to reduce him to helplessness, and to make him liable to injury when he does finally use these tools.

Herbart and His Work.—While Pestalozzi was at Burgdorf, Johann Herbart (1776–1841) was acting as tutor to the three sons of the governor of Interlaken. Herbart's mother saw to it that he left school before the completion of his course, because she believed that association and experience in the world was far better for him than the formal and speculative philosophy of the classroom.

Herbart came into contact with Pestalozzi and, absorbing the enthusiasm of the elder man, accepted many of the ideas of the experimenter at Burgdorf, and rejected much of what he considered impractical. He maintained, in his "Umriss pädagogischer Vorlesungen" (quoted in Bennett), that every boy should learn to handle the recognized tools of the carpenter.

Elementary schools, he further states, should have workshops, and every child should learn to use his hands.

Froebel and Self-Activity.—Perhaps the most direct heir to the doctrines of Pestalozzi was Wilhelm Froebel (1783–1852). He firmly advanced the doctrine of self-activity, which became the keystone of his educational theory. With a varied career in school and out of school, Froebel did not enter the field of education until 1805, when he first became a teacher. During the next two years he studied the work of Pestalozzi so that he might carry on the great work of his idol.

In 1816 he took charge of the education of his niece and four nephews, and with this group formed the nucleus of his educational colony. In 1826 he published his "Education of Man," in which he stresses the high place of handwork in the process of education. He states that the activity of the senses and limbs of the infant is the first stage of education; play, building, modeling are the important stages in the development. He further advocates a two- or three-hour period each day in which the boy devotes himself to the production of some definite external piece of work. Lessons, he says, through and by work are vastly more impressive and lasting than lessons from books. It predicts that schools will eventually come to an arrangement whereby provision is made for work along with study. W. H. Hailmann, who translated the "Education of Man" into English, lists the kinds of work suggested by Froebel under the headings of work in the field, garden, forest, and house. His list of occupations includes the preparation of fuel, weaving, binding of books, woodwork, care of the garden, making of tools and baskets, painting, and a host of other needful things.

Froebel's plan for carrying out this program failed. It served, however, as a guide for those who came after him. His contribution to kindergarten education was his great work. Here he put into practice his ideas of handwork.

Wehrli and the Training of Teachers.—Wehrli, the associate of Fellenberg, in 1833 accepted the management of a normal school at Kreuitzlingen on Lake Constance. Here he put into practice his theory of training rural school teachers by combining the usual school studies with farm work and instruction in the art of teaching. He firmly believed that the teachers of the children of the poor must be friends and associates of the poor, yet well educated. He lamented that so many wealthy young men, who had lived luxuriously at college, became teachers in the peasant schools. To avoid this evil, he attempted to make the student's life simple and humble. At Kreuitzlingen, a large farm, worked by the students, assisted in the support of the institution. A chart of the course of instruction is printed in Henry Barnard's "National Education in Europe" (reprinted in Bennett, page 168). It indicated how work was varied with study in the school day which started at five o'clock in the morning and ended at nine o'clock in the evening.

Many agricultural schools grew up in Switzerland due to the impulse imparted by the Hofwyl experiment. These came to be known as Wehrli schools. They were for the sons of farmers who had completed the elementary school. At these schools were taught the science of agriculture, chemistry, mathematics, geography, languages, and history. A farm well stocked with cattle, gardens, and farm requisites was attached to each school. Each day, for about five hours, a competent farmer instructed the boys in practical farming. The produce of the farm was supposed to cover the cost of instruction. Perhaps to these schools, which were established in every canton, can be laid the superiority of the Swiss farms, which yielded a greater return for the outlay than any other farms in Europe.

Oberlin and the Infant School.—In 1766, before the work of Pestalozzi and Fellenberg, France produced an educator whose influence upon future educational theory is worth noting. This man was John Oberlin (1740–1826). In a small community he established an infant school, probably the first, which served as a model for the "écoles maternelles." Robert Owen visited it after he had established the first British infant school. It was in origin some twenty years earlier than Froebel's kindergarten.

This institution was based upon the theory that the child from the cradle was capable of being taught to distinguish right from wrong, and of being trained in the habits of industry. Instruction was mingled with entertainment. Two women were in charge, one to teach the handwork, the other to entertain.

The Industrial Revolution in England and the Work of Robert Owen.—The application of steam to industry and the invention of the power loom made possible in England the use of a larger number of laborers with less strength and skill in the making of cloth. Children were ideal tenders of machinery which required merely watching and setting. Droves of children entered industry, and the workhouses were only too glad to play into the hands of the employers in sending them vast numbers of children. A system which employed children as young as six years for twelve to fifteen hours a day was monstrous, to say the least. In 1796, due to an epidemic, the Manchester Board of Health pointed out the evils of the system. Agitation against the evils of the factory system was not slow in getting under way, and, by 1802, Parliament acted. The result was the Factory Act of 1802. This act limited the hours of labor to twelve, between 6 A. M. and 9 P. M. It further required that instruction in the three R's should be given to apprentices.

In the discussions leading up to this law, and in factory affairs in general, a major part was taken by Robert Owen (1771–1858), a successful factory manager and the founder of English socialism. In 1799 he became manager of a cotton mill at New Lanark near Glasgow, Scotland. Here he put into practice his ideas on social theory. In a speech to the not-eager community he anticipated the behaviorists by some one hundred twenty years in stating that man becomes a wild, ferocious savage or a cultured gentleman according to the circumstances in which he may be placed from birth. From this he argued for the education of the poor, and demanded for them a place in the schools.

Owen's School at New Lanark.—In 1816 he established an infant school at New Lanark. The school was carried on in the Pestalozzian manner under the direction of one James Buchanan, an illiterate, selected because he knew nature and loved children. Great work was done for the children of the factory, and much was learned by educators who visited New Lanark.

Owen's Venture in America.—Persecuted in Britain, Owen embarked for America in 1825, to create a new moral world. He purchased 20,000 acres in Indiana and named it New Harmony. William Maclure, another wealthy man and a Philadelphian, associated himself with Owen in the venture. Maclure had traveled widely in Europe and had visited New Lanark, Yverdun, Hofwyl and other centers of educational experimentation. Fellenberg he considered autocratic, while Pestalozzi, he noted, was the essence of democracy. Maclure attempted to secure the services of Pestalozzi in order to set up a school in Philadelphia. He failed to lure the master to America, but succeeded in engaging one of his assistants, Joseph Neef. The first Pestalozzian school was opened in 1809 in Philadelphia. Neef, upon the failure of the attempt some years later, joined Owen and Maclure in the New Harmony venture.

Owen, in seeking a new moral world, wished to create and develop a new social organization which would rationally educate and employ everybody. Maclure wished to make New Harmony the educational center of America, through the introduction of Pestalozzian system of instruction. Neither of these aims was accomplished. But the influence of the attempt was widespread. Many eminent scientists visited the institution, and imbibed the spirit of Pestalozzi and Fellenberg and saw in operation the theory of the equal education for both sexes. The infant school was based directly upon that at New Lanark, and the higher school, under the direction of Neef, upon the evening school at the New Lanark mills. The instructors had been pupils of Pestalozzi.

Silliman's Journal (mentioned in Bennett, page 177) in 1826 printed an outline of the course of study. It indicated that the teaching was facilitated by the study of models, largely of the students' own creation. Hygiene, for example, was taught by experience and observation rather than by philosophical methods. Lithographing, engraving, and printing, as well as other mechanic arts, were in the curriculum. An adult school, for those over twelve years of age, was conducted in the evening for those who wished to continue their studies in mathematics and the useful arts.

The second school was the real Pestalozzian school, and to it Maclure devoted his energy. He believed that every child of the productive class should be taught a trade in order that he might be self supporting and independent. This trade was to be selected by the child himself, who, if unable to select any particular calling, was to be assigned to a trade.

After two years, due to differences of opinions among the leaders, the community broke up, and Owen returned to Scotland. Maclure sought to develop new educational projects from the wreck, but all his attempts were failures.

George Lockwood in his "The New Harmony Movement," page 261, gives this quotation from Maclure's "Opinions":

"Being taught to make shoes or coats does not force the possessor of such knowledge to be a shoemaker or a tailor, any more than learning mensuration or navigation obliges one to become a surveyor or sailor. They are all acquirements good to have in case of necessity. . . . Even at present, our farmers and manufacturers, nine-tenths of our population, would be very much benefited by possessing one or two of the mechanic arts suitable to their occupations."

Alcott and Self-Education.—Amos Bronson Alcott (1799-1888) applied Pestalozzian principles in his school at Cheshire, Connecticut, at the time when Owen was endeavoring to work out the New Harmony experiment. Conservative New England found Alcott's educational ideas far too radical, and consequently his work was not popular. After many years spent in theoretical explanations of his ideas, he opened his school in the Masonic Temple, Boston, in 1834. After a few years his advanced religious ideas and his having admitted a colored child reduced the student body to five pupils. He closed the school in 1839 to retire to Concord, where he found solace in the society of Emerson, Thoreau, and Hawthorne.

Alcott believed in education as "a leading of the young mind to self-education." All subjects were considered as developers of the spiritual faculties, the imaginative faculties, or the rational faculties. Writing and sketching from nature found a large place in his school. Cultivation of plants and the development of environmental happiness through pleasant and beautiful surroundings also were paramount features of the Masonic Temple school.

The Manual Labor Movement in America.—Fellenberg's academy gave impetus to the manual labor movement in America. This movement began about 1825, reached its apex about 1835, and waned, as an educational force, about 1845.

In America the Fellenberg idea of class education was not carried out, and the farm and trade school were combined with the academy. Manual labor, then, had two purposes—the physical (and allied mental) training, and the productive or monetary return. Young men, the advocates of this educational method pointed out, could secure an education at a minimum cost and also enjoy the benefits of physical exercise. The need for exercise, and also for money, was greatest among theological students, and hence these institutions took the lead in the movement.

The American Education Society was devoted to the promotion of ministerial education, and through its publication, "The Quarterly Register and Journal," espoused the cause of manual labor. The Reverend Elias Cornelius (1794–1832) was the secretary of the society, and was regarded as the leading spirit of the movement. Cornelius in 1829 published in Volume II, No. 2 of the Register, page 108, a summary of the movement up to that time and also the answers to a questionnaire which he had sent to schools in various parts of the country which had adopted the manual labor idea. In 1830 W. C. Woodbridge published in the American Annals of Education his letters entitled "Sketches of Hofwyl."

Cornelius termed most successful the experiment at the Andover Theological Seminary, Andover, Massachusetts. The movement here was started in 1826 by a few individuals solely for the purpose of invigorating and preserving their health. The trustees saw the pecuniary value of the idea, and erected a shop fully equipped with benches and tools. The work done consisted in boxmaking of various sorts and other carpentry projects. One and one-half hours a day was given to this shop work. By 1829 the profits from production had been sufficient to defray the cost of equipment and to cover the wages of two mechanics.

Another interesting experiment during this period was that at the Maine Wesleyan Seminary, founded in 1825. This was a preparatory school for three types of young men, (1) those "worthy poor" who wanted an education, (2) the idle well-to-do who needed proper stimulation to interest them in work so as to keep them from dissipation, and (3) especially talented students who "for want of some regular and systematic exercise, were doomed to find an early grave." (Annals of Education, Edited W. C. Woodbridge, Boston. Chapter 1, page 82.)

In connection with the seminary, a 140 acre farm and a mechanical shop offered possibilities for manual labor. The shop was equipped for chairmaking, cabinet work, and tool making. An unsuccessful attempt to institute coopering and shoemaking occurred early in the experiment. The *Register and Journal* of the American Education Society gives (Vol. II, page 110) the report of the academy. Forty-two students paid their board and tuition without labor, while sixty-five were in the "laboring department." The advantages of the plan are set forth in the same report. The three year course offered was divided into two sections; one was purely college preparatory, the other was designed for those who were "willing to labor with a knowledge of agriculture or one of the mechanic arts."

The Oneida Institute of Science and Industry at Whitesborough, New York, was another signal attempt in the Fellenberg manner. This was opened by the Reverend George Gale in 1827, a minister who had lost his health through too close application to study. He retired to a small farm to recuperate, and, upon rapidly returning to health, came to the conclusion that vigorous exercise was the remedy for his trouble. To experiment upon this thesis he offered to board and instruct eight young men in college preparatory work if they would agree to work three and one half hours a day in the fields with him. His report in the *Quarterly Register and Journal* of the American Educational Society for 1829 indicates that the experiment was satisfactory as regards students' health, educational progress, and production of farm products.

The results of this experiment, when laid before the Oneida Presbytery, created a movement to widen the scope and enlarge the facilities of the Reverend Mr. Gale's establishment. The result was the Oneida Institute of Science and Industry. This was located on a 114 acre farm. The trustees equipped the farm fully, and the instructors worked it. Student labor could be employed, providing only three or four hours of it were utilized each day. The instructors were to have the farm produce, but it was required that they board the students and instruct them. They received fifty cents a week in addition to this. The first year was successful, and a profit was netted. Unfortunately a flood ruined the crops the second year, and the Institute worked under a deficit. The third year was also a success. In the "Proceedings of a Meeting Held at the Masonic Hall on the Subject of Manual Labor in Connection with Literary Institutions," New York, 1831, it is noted that the Oneida Institute had become so popular that 500 applicants were turned away that year, since the accommodations provided were limited to 60. (Bennett, pages 186–188.)

The Reverend John Monteith established the Manual Labor Academy of Pennsylvania at Germantown, a direct outgrowth of the Oneida Institute. At first, when Monteith proposed his plan to the brethren, it met with a skeptical reception. It was termed incompatible with a student's life and absolutely unsanctioned by example of past experience. However, in 1828, $350 was appropriated for the rental of a building, and four students joined Mr. Monteith. Within a year the roll had increased to twenty-five (reported in First Annual Report of Board of Trustees, Manual Labor Academy of Pennsylvania— Philadelphia, 1829). Carpentry, gardening, and farming constituted the student's labor. Several of the pupils were good workmen in wood, and these acted as instructors to the less experienced. The report states that these were profitable in their own labor. Germantown, then, was perhaps the best example at that time of instruction in woodworking. Up to this time, manual labor schools had stressed in their reports the physical-exercise aspect of the work or the pecuniary side as a means of support. Nothing is said of woodworking taught by skilled workers until the Germantown school report.

At a meeting of a group interested in manual labor in New York City, on June 15, 1831, the question of the introduction of manual labor into literary institutions was brought to a focus. Papers were read upon various aspects of the question, more especially stressing the physical benefits and monetary returns to be gained. A committee was appointed to consider the whole question and to promote the movement as it saw fit. In July of the same year the committee called another meeting, when the Society for Promoting Manual Labor in Literary Institutions was founded. Theodore Weld, of the Oneida Institute, was its appointed head.

Weld travelled about the country, visited institutions, spoke before assemblies of educators, and collected data which he intended to present before the society. An accident while crossing a stream in Ohio befell him, and he lost his records in the mishap and barely escaped with his life. Thus, when he appeared before the society, his plea was based solely upon opinion, not upon statistical data. He asked for the introduction of manual labor for physical training purposes—the training of mind and body to reciprocal action. He showed that one of the great defects of the educational system was the lack of provision for health education. Bennett, in the "History of Manual and Industrial Education up to 1870," quotes freely from Weld's "First Annual Report of the Society for Promoting Manual Labor in Literary Institutions." He notes the reasons advanced for the manual labor innovation. After giving his report, Weld resigned, and the Manual Labor movement declined. The failure, for obvious reasons, of the institutions to be financially successful with the sporadic labor of a few students seems to be the cause for its marked decline after 1850.

The Development Since 1870

Manufacturing Replaced Agriculture as the Leading Industry of the United States After the Middle of the Nineteenth Century.—The effects of the Industrial Revolution manifested themselves pointedly in the economic, social, and moral life of the American people only after the middle of the nineteenth century. Then it was that the industrial phase of the American social structure supplanted the agricultural and commercial structure of colonial days. The development of big business has been noted in Part One of this volume. It remains to trace the roots of the educational development.

Types of Vocational Schools.—When the field of industrial education is surveyed, it can be broken up into several divisions. These divisions are based upon type schools. Bulletin 17 of the Federal Board for Vocational Education lists six types of trade or industrial schools or classes which may be organized under the provisions of the Smith-Hughes Act:

    A. Unit trade schools.
    B. General industrial schools in cities under 25,000 in population.
    C. Part-time trade extension schools.
    D. Part-time trade preparatory schools.
    E. Part-time continuation schools.
    F. Evening industrial schools.

The same bulletin in great detail sets forth the controlling purpose of each type of school; together with information regarding admission, hours, entrance requirements, programs of studies, methods, and qualifications of teachers.

To these types of schools listed officially by the Federal Board there may be added the full-time trade school, the corporation school, the correspondence school, the girls' vocational school, the industrial arts division, the cooperative school, the engineering or technical school, and modern apprenticeship systems in the industries. Mention might also be made here of the numerous foremanship training schemes, round tables, and discussion groups.

The Roots of These Schools.—As has been shown, many of these schools had their origins in remote antiquity, not a few in the middle ages, and many in the seventeenth and eighteenth centuries. The great development and enrichment in the field of vocational education, especially industrial education, demands an explanation in terms of origin for purposes of true understanding and as a basis of future action.

As the pioneer and the frontiersman disappeared, and the forests gave way to cities, the restless spirit could devote himself to internal developments. Industry gives evidence of this in an unquestionable manner. Not a few attribute educational development also to this fact.

The Civil War Closes the Era of Apprenticeship Training.—It was not until after the Civil War period that it was deemed necessary to develop a new system of education to take the place of the apprenticeship system. Before this time the emphasis had been toward bolstering up the old system of apprentice training.

Many Types of Schools Necessary to Solve the Problem of Industrial Education.—The diverse systems of industrial training, together with the numerous types of industrial schools, give evidence that no one type of training is adequate for industrial training. The specialization in industry and the breaking down of the job into its component parts has thus reflected itself in the schools.

Arthur Mays, in "The Problem of Industrial Education," page 7, notes the difficulties encountered by the forces of production and the forces of education. He notes the problems of the joint management of industrial training by school and industry. Then he makes this significant statement: "It now seems probable that the public schools will be the chief school agency in industrial education."

The parties concerned with industrial education are: the

school, the young worker, and industry. Society is also vitally interested, for these are some of the factors of society.

The Problem of the Training of the Worker More Acute in the Modern Era.—Since 1870 industrial history has shown greater and greater efficiency in the development of manufacturing. Processes, methods, machines have been steadily perfected. The industrial engineer has become a necessity. One phase of the subject has been largely neglected. That is the guidance and education of trained workers.

Leisurely craftsmanship disappeared after the effects of the Industrial Revolution became manifest. Local industries gave way to national industries competing for world markets. Quantity production supplanted small production. The artisan gave way to the machine. The journeyman no longer owned his tools, and his dependence upon a capitalist became fixed.

The New Type Worker.—The new type of industrial worker of America appeared as a sizable group only after 1870. Now this type dominates the whole industrial field. John A. Fitch in the Pittsburg Survey by the Russell Sage Foundation in 1911 (page 349) indicated that, of 23,337 employees of the Carnegie Steel Company plants in Allegheny County, 83 per cent were low grade workers. Paul Douglas in his study of "American Apprenticeship and Industrial Education" done at Columbia University in 1916 (page 119) showed that 72.3 per cent of the workers in 189 plants in Chicago were unskilled and low-grade skilled workers. The unskilled group included, among others, women, children, recent immigrants, and non-vocationally trained adult males. As a group these stood very low in the social scale, displaying the customary shortcomings and vices of their social position. In the group will be found the shiftless, the inefficient, the drifters, and the seasonal workers. Their problems can be met by compulsory school laws, continuation schools, and a "pusher" trade training.

The skilled worker constitutes a similar problem, and his need for industrial education is as pressing as that of the low-grade or semi-skilled worker. The novice who has not entered industry is another type for whom industrial education is essential. Then follows the technical expert, the foreman, and the assistant foreman. The myth that machine industry has reduced the proportion of skilled workers is ably discussed pro and con in such works as Arthur Pound's "The Iron Man in Industry" (*The Atlantic Monthly Press*, 1922) and *The Executives Magazine* (reprinted in the *Literary Digest*, Feb. 7, 1925, page 24). In Prosser and Allen's "Vocational Education in a Democracy" a whole chapter is devoted to the place of the non-automaton in industry.

Specific Trade Training Necessary.—It has been noted that through the early development of modern industrial education, the idea was prevalent that a general and fundamental training should be given to embrace all the trades. It was only later that the idea of specific trade training found room in this democracy. The objection raised to trade training was that it cut the young person off from all possible advantages of a liberal education and from the possibilities of advancement to higher occupational levels. The realization that those who enter the trades usually have not the chance of entering a cultural course or of remaining in it was slow in making itself felt.

The emphasis on "broad" education and the fear of specific trade training agencies hindered the development of the full-time unit-trade schools.

Manual Training and the Faculty Psychology.—Manual training was one of the plans utilized to give the desired trade preparation. It was defended in terms of a faculty psychology, and it rested upon the idea of the transfer of training. The failure of the manual training idea, due to its origin in faculty psychology, caused a change in the thought on trade training.

The Era of the Private Trade School.—It has been noted that during the middle of the nineteenth century private trade schools were established to "elevate" the working classes. The public schools, the educators said, were for basic or general education. The private schools arose then in defense of specific trade training.

The early trade schools were organized on a single trade basis. The building trades and the metal trades were invariably represented. Arthur Mays in his "The Problem of Industrial Education," page 100, says of them:

> "Many of the schools were well organized, and splendidly equipped, and accomplished excellent results. Nearly all of them are still in operation, rendering a most valuable service to the relatively small number of young men and women who can profit from this type of vocational education."

Among the pioneer trade schools were the New York Trade School (1881), The Philadelphia Builder's Exchange (1890), the Baron de Hirsch Trade School (1891), Drexel Institute (1892), The Clara de Hirsch Trade School for Girls (1899), and the Manhattan Trade School for Girls (1901). Since 1907 the city systems have taken over several of the private schools, while New York City has founded the Vocational School for Boys, (1909), the Murray Hill Vocational School (1914) and the Brooklyn Vocational School for Boys (1915).

The Milwaukee School of Trades (1906) was the first school to be supported by a tax levied specifically for industrial education by a municipality.

Surveys of the Problem of Industrial Education After 1900.—A movement for scientific determination of the problem of industrial education can be noticed after the year 1900. The Douglas Commission on Industrial Education was organized and made its comprehensive report. This is the well known "Report of the Massachusetts Commission on Industrial and Technical Education, 1906." Arthur Mays sums up the report in his "Problem of Industrial Education" (page 102):

> "This report reviewed the efforts of manual training and all of the other existing types of industrial work in the schools. It pointed out the failure of manual training as a means of meeting the need for trained mechanics, and praised the work of the trade schools."

The United States Commissioner of Labor undertook surveys of industrial education in Europe and America. The annual reports of this department of the government for 1902 and 1910 are devoted solely to industrial education.

The National Society for the Promotion of Industrial Education.—The year 1907 saw the creation of the National Society for the Promotion of Industrial Education. This society was composed of outstanding men from the fields of industry, education, and politics. It proved itself a most effective agency in promoting vocational education. Among the important surveys it has promulgated are "The Vocational Education Survey of Richmond, Virginia, 1914" and "The Minneapolis Survey for Vocational Education, 1915." Later the work of the American Vocational Association also proved of great importance. Those important groups and works contributed largely to the legislative enactments, the most important of which has been the Smith-Hughes Act, to be described in the section on legislation.[3]

The Burden of Vocational Training Shifted to the Public Schools.—The tendency, then, appears to be toward a scientific determination of needs, followed by a scientific construction of a plan to meet these needs. This attitude characterizes the trend in vocational education for the opening of the twentieth century. To this must be added the shifting of the burden of trade education to the public school. This shifting was, to be sure, received reluctantly by the public school.

The Limitations of the Private Trade School.—One of the reasons for the apparent lull in the trade school movement and in their establishment can be explained by the fact that the need for skilled workers, though great, is small in relation to the need for semi-skilled and unskilled workers. Trade schools are schools of secondary grade; the number of children who can pursue the trade school course is limited, and of this group few enter the mechanical trades. Many students attend trade schools until the time when they secure positions, and then they leave school. The regular high school draws the greater number of the individuals who can pursue secondary work and are financially able to remain at school. The difficulties in making

---

[3] For the full history of the National Society for the Promotion of Industrial Education, which later became known as the American Vocational Association, the reader is referred to Dr. Struck's book: "The Foundations of Industrial Education." (John Wiley & Sons, 1930.)

occupational choices also limit the number which might enter trade schools. Still another reason is the high per-capita cost of trade school education. This is due to the necessity of having modern equipment and much material. The selling of manufactured articles by the trade school usually is not sufficient to cover the cost of production. High tuition charges are necessary in the private trade school, higher taxes for the public trade school.

The trade school, with its limitation, has not developed as a super-apprenticeship system. In times of labor shortage there is a demand for the trade school by non-educative bodies, but the place of the trade school in the structure of industrial education is limited.

The Factory Schools Arose to Meet a Definite Need.— The extensive specialization in industry which has followed the Industrial Revolution has been noted. Mention has also been made of the intricate machinery which has been devised for industrial operations and processes. Indeed those two factors of production are those largely responsible for the overthrow of the apprenticeship system. Their influence on industrial education has been positive as well as negative. The origin of many a corporation school has been due to these two factors.

It is, therefore, not surprising to find the factories of America attempting instruction of a formal sort within the factory close. The current educational philosophy backed up the factory in its endeavor to give highly specialized training.

One of the first of the factory schools was that of Hoe and Company, of New York City. This large corporation manufactures printing presses. A plant school was established in 1872, to train apprentices in the skilled trades, especially the machinist trade.

Other plant schools followed, including those of the Westinghouse Electric Company in 1888, the General Electric Company and Baldwin Locomotive Works in 1901, and the International Harvester Company in 1907.

After 1905, a general impetus is noted in the establishment of factory schools. By 1920 most of the large manufacturing and service corporations had schools. Such names as Western Electric Company, Goodyear Tire and Rubber Company, Ford Motor Company, The Morton Company, Mergenthaler Linotype Company, National Cash Register, Brown and Sharpe, Yale and Towne, and so on, merely give an idea of the representative schools in diverse fields functioning at the present time.

These schools have arisen to meet a definite need, a need growing out of the industrial world in which they function. The training of skilled mechanics and specialized workers has been the major objective of these institutions.

The Limitations of the Corporation School.—The limitations of the corporation school are worthy of mention. Only the large corporations seem to have plant schools. Morris in his "Employee Training," page 191 (quoted from Mays, "The Problem of Industrial Education," page 128) finds that in plants employing less than 800 only very limited programs of employee training can be carried on successfully. On page 291 he states: "This investigation has, however, found plants which employed a working force of not over 3,000 people, operating successful educational programs." The inability of the corporation to hold the graduates of their shop schools is another limitation which has caused the directors of such schools much worry. Since industry and the public school are the major agencies interested in trade training, some form of corporation school appears to be part of the solution of the problem of industrial education. Just what use educators will make of industrial shops as school laboratories, and just what amount of supervision on the part of educators will be required for students in industry, is not yet determined. However, since the root of the problem is being nourished by both industry and the public school, the solution will be a composite device utilizing the experiences of both parties.

The Correspondence School.—Any study of the means and devices of industrial education must include a mention of the correspondence school. The general reaching out for education, together with the real need for trade education and the failure or inadequacy of existing educational agencies, brought forth the correspondence school. The numbers enrolled, their progress, their lack of progress, and the reasons for success or failure are usually secrets stored only in the archives of the school. John Van Liew Morris, in his "Employee Training" (1921), page 241, makes this statement, after noting the difficulties in culling statistics on this phase of the subject: "Probably more men in American industry have gained the technical phases of their trades from correspondence schools than by any other means."

There are two types of correspondence school—the privately owned type, which is run primarily for the profit it can make, and the tax-supported college or university and the endowed school.

Clever advertisements, indicating promotion, greater pay and the like, are factors in the reasons for the large enrollments of private correspondence schools. It is unnecessary to note here the limitations of educational work without the personal guidance of a teacher, the tremendous student mortality, and the other hindrances and shortcomings of correspondence courses.

The Evening School.—The origin of the evening school is not difficult to explain. The idea of studying in spare time was fundamental to all ideas upon trade education. The economic utilization of evenings appealed to industrialists, educators, and pupils.

The Dutch Colony of New Netherlands, according to Robert Seybolt ("The Evening School in Colonial America," University of Illinois, 1925), had evening schools prior to 1674. Seybolt quotes (page 10) a clause of the New York indenture provisions of 1701—"in the evenings to go to school each winter to the end he may be taught to write and read." Arthur Mays after studying the problem arrives at this conclusion (page 146 of "The Problem of Industrial Education"):

"That evening schools were quite common throughout the Colonial period seems clear, and it is evident that they were chiefly vocational in purpose, supplementing the daily work of apprentices and others engaged in the customary occupations of the times."

Society accepted the evening school as a seemly mode of

vocational education. It has been accepted as an economic means of vocational and general education, and, hence, it has prospered and developed until it occupies a prominent place in the scheme of American education.

The position of the evening school is shown in the "Fourteenth Annual Report to Congress of the Federal Board for Vocational Education" (page 3).

The enrollment in federally approved schools is given in Table I. The evening school enrollment for the year ended June 1930 is given as 341,565. This is sufficient to enable one to see clearly the growing interest of the worker in vocational education in the evening school. The recognition of the fact that the worker needs a broader technical knowledge than can be learned by the "pick-up" method is the fundamental root of the evening school movement in trade education.

The Structure of Modern Industry Precludes the Possibility of Great Use of the Full-Time Vocational School.—When education was for the few, the elect and the chosen, the full time school met the needs. With the coming of the factory system and the urbanization of the population some form of part-time education for those who must work was necessary. Schools were established to permit those who came to them to continue their education while pursuing gainful employment. The origin, then, of the continuation school is accounted for. It is a normal development and an outgrowth of the modern factory conditions and the economic status of modern life. As such, it is, without doubt, an institution which can be adapted to modern educational needs.

Part-Time and Continuation Schools.—The growing desire to maintain school control over children during their plastic years, and the recognition of the futility of traditional secondary school training for the vast majority of adolescents are further causes of the establishment of part-time and continuation schools.

The part-time scheme is an attempt at the solution of the problem of how to fit education into the modern industrial world. Education in a democracy cannot be for the few, nor can it be reserved solely for the leaders. Schools must be set up for those whose early participation in industry is a necessity. The great industrial nations of the world have experimented with the continuation school as a necessary feature of their educational programs.

In America there have been two types of part-time schools. One type has sought to assist individuals between the ages of sixteen and eighteen, while the second has dealt with those from fourteen to sixteen. The former is usually known as the "part-time trade extension" school; the second, as the "general continuation school."

The youth from sixteen on is eligible to begin some sort of specific trade training. Hence the part-time school seeks to give him training along a specific line. The aim has been to extend and supplement the specific trade or occupational activities which the youth experiences each day in his business activities.

The second type of continuation school training, that for the fourteen-year-olders, is quite different in its emphasis from the first type. The aim of this type of school has been to provide for the continuation of the education of these children who have left school to find employment. Guidance and training are the characteristics of this general continuation school education. It has sought to utilize the daily activities and experiences of the child in order to help him find his place in society. The home life and the occupational life of these children are studied in solving the problem of his adjustment. These schools seek to supply their pupils with skills, facts and attitudes that will guide them in their daily life. Reading, writing, and calculating are among the first skills taught these pupils. Manual and industrial skills come next to these fundamentals. Occupational courses of many kinds are provided, and the pupil is often allowed to sample each. The student's present needs are carefully noted, and his time is not wasted on non-essentials or traditional exercises. Character development and poise have found ample place in the activities of these schools.

A Statistical Summary.—A few figures from the Thirteenth Annual Report to Congress of the Federal Board for Vocational

Education, page 63, serve to indicate graphically the throngs which are taking advantage of part-time vocational education.

In Trade and Industrial Schools

Number of pupils enrolled in vocational courses

Year	Evening		Part time				All day	
			Trade extension		General continuation			
	Male	Female	Male	Female	Male	Female	Male	Female
1929	123,503	7,581	28,468	7,693	168,120	163,293	56,537	8,301
1928	106,872	7,757	33,656	8,875	162,798	160,214	49,317	8,122
1927	90,923	6,651	32,591	6,815	163,790	144,743	43,204	6,912
1926	82,863	6,831	30,640	11,194	150,906	139,452	39,346	5,453

These figures refer to vocational courses which are Federally aided.

The Industrial Arts.—The so-called "industrial arts" have found a unique place in the scheme of American education. It is necessary to trace the origin and growth of this feature of the general education program which relates so satisfactorily to the vocational education program.

The Forerunners of the Industrial Arts.—Two major factors appear as precursors of the industrial arts curriculum. The first is the demand for industrial training so noticeable after the Civil War—the demand, in other words, for trained mechanics. And the second, the demand on the part of certain educators for a liberalized school curriculum. This demand grew out of a realization of the democratic status of education. It was a demand for an education suitable for the children of all classes. It was a demand to train the "whole boy" and not only his head.

This movement gained ground slowly during the eighteen-seventies, despite the hostile philosophy which considered specific

INDUSTRIAL ARTS

trade training impossible as a public school activity. It was necessary, then, for the liberals to seek some form of training which would be at the same time industrial in content and yet general and abstract—or, as they were wont to say, "disciplinary." This training was to train the "general faculties."

Two men stand out as leaders in this movement according to Arthur Mays ("The Problem of Industrial Education," page 191): Dr. John Runkle, president of the Massachusetts Institute of Technology, and Professor Calvin Woodward of Washington University, St. Louis. The former was looking for a scheme whereby he could teach the processes and mechanical principles of industry to engineering students; the latter advocated teaching all boys shop work in order to enrich their education, regardless of future vocation.

"Manual Training."—Runkle in looking about for a course of study sufficiently general and applicable to traditional school room procedure happened upon the exhibit of the Imperial Technical School of Moscow at the Centennial Exposition at Philadelphia in 1876. Here, in the set of show exercises, he found much material that he could utilize—in fact a solution to his problem of practical mechanics for engineers. The following year the Massachusetts Institute of Technology established a series of mechanic arts shops. Woodward at once adopted the Runkle plan, and made handwork a part of general education. He called the new work "manual training." It is due to him, perhaps, that manual training became an indispensable part of the public school curriculum.

In 1880 the pioneer secondary school which emphasized manual training was opened. This was the St. Louis Manual Training School at Washington University. Woodward here gave to manual training its stamp—general training as opposed to specific vocational training.

Industrial Arts Different from Manual Training.—Modern industrial arts training is different from the manual training of Runkle or Woodward. As the faculty psychology was superseded, so the idea of the disciplinary value of manual training passed away. For uninteresting exercises with material there were substituted real, useful, and interesting problems which involved use of tools and processes necessary to the doing of a genuine job.

The Sloyd Influence.—At this time it is necessary to note an influence from Sweden. That was the influence of Otto Solomon with his series of graded models designed for household use. The term Sloyd was applied to these exercises. Although many of these exercises did not apply to pertinent problems, the influence of the movement gave a characteristic touch to the development of industrial courses in the grade schools.

Manual Training Failed to Prepare for Efficient Participation in Industry.—After 1900 there came a demand in education for real training for industry. The twenty years of manual training had produced no skilled operators. The manual training exponents had held that, in addition to general education, manual training had an important part in the preparation of boys for industrial positions. To vindicate their programs, a renewed interest was taken in the subject, and the result was an enlarged and enriched program of industrial arts education. This program was different in content and method from that which had preceded it. School shops took on factory plans and layouts, more and better machinery was installed, jobs became more interesting and of greater worth, and a more reasonable tie-up was made with facts and information relating to industry. Although the manual arts people failed to demonstrate all that they claimed for manual training, they created, at least, sufficient disturbance to promulgate thought on the subject of trade training.

The Cooperative Plan an Attempt to Correlate Learning and Earning.—The cooperative system of trade training has grown out of the attempt to correlate trade experience with industrial education. The cooperative system is based upon an agreement between a manufacturer or group of manufacturers and a school system, whereby the manufacturers agree to give both general and specialized instruction to the apprentices. The work in the shop is formulated by the shop with the approval

MERITS AND WEAKNESSES

of the school authorities, while the work in the school is arranged by the school authorities.

The apprentices are usually divided into two groups, so that while one group is at work in the shop, the other is in the classroom. Hence, shop and school are always occupied.

The plan recognized the fact that one may be educated by his work as well as for his work. It also recognized that the desirable combination of work and study, which was formerly possible for large numbers of children and youths, is impossible today without a carefully planned scheme of cooperation between the employers and the schools. There is nothing more unfortunate in the present social order than the necessity which confronts so many young people today of choosing between all school and all work. This choice the cooperative system makes unnecessary, because, while he is in the shop, the student is a real wage earner.

Dean Schneider the Pioneer.—Dean Schneider, of Cincinnati, was the pioneer of the cooperative system. He established this system at the College of Engineering of the University of Cincinnati during the first decade of the twentieth century.

At Cincinnati a student begins his work in the foundry, passes to the machine shop, then works in the assembly department, and next goes to testing and inspection divisions. Toward the end of his course he finds himself in the drafting rooms and executive offices. In short, he follows the natural flow chart of the product, from the raw material.

In the shop the boy is paid the prevailing wage of the shop workers.

Colleges of Engineering Adopt the Cooperative Plan.—Similar plans have been worked out at Harvard, Antioch College, and the Massachusetts Institute of Technology. In 1923 the industrial engineering department of New York University instituted the cooperative system, and in 1925 all the departments of the College of Engineering of that institution adopted the cooperative system.

Merits and Weaknesses of the Cooperative Plan.—The cooperative movement may be considered as a step in the solution of scientific training for the industries. The cooperative plan can be undertaken with very little initial cost when industry furnishes the laboratories. It can be maintained at a relatively small cost, and is capable of adaptability to the needs of individual students. It is a system that is in harmony with the trend toward specialization in industry. The cooperative school with shop practice under factory conditions is more practical, in most instances, than the trade school, which depends upon its own shops. The cooperative school does not run the risk that is so often incurred by the corporation school of being narrow and of subordinating all other interests to the interest in immediate profits. The cooperative school can be carried on efficiently, and, at the same time, employ the latest and best standards of pedagogy, equipment, and administration.

The weakness of the cooperative system lies in the possibility of poor coordination of the shop work and class work. It is the function of the coordinator to correlate the work of the two branches of the student's activity. This teacher must be in constant touch with the student when he is at work in industry. This contact can be obtained by weekly reports of the student to the coordinator and by frequent visits by the coordinator to the pupil in the shop. It is obvious that a coordinator who is laden with administrative and clerical duties will prove of little value.

Much attention has been given to a cooperative plan for apprentice training developed by the metal trades of Milwaukee. Here the metal factories cooperate with the public schools. A director of apprentices, employed by the metal trades organization, coordinates the work between the shops and the schools. Part of the student's time is spent in the industry, and the rest in the school. The school period is further divided between classroom and school shop.

This plan allows the small industry which is incapable of conducting a full training program to reap some of the benefits of trade training. It also allows the school the use of modern equipment and training on real jobs. Pioneer schools employing the cooperative plan include the Fitchburg High School, Fitchburg, Massachusetts, the Beverly Industrial School of Massachusetts, Technical High School, Providence, Rhode Island, and the Cincinnati High School.

The Spread of the Cooperative Schools.—Cooperative vocational schools are now common in a number of states, including Ohio, Pennsylvania, Maryland, California, and Massachusetts. The Federal Board for Vocational Education has prepared a special bulletin upon this phase of industrial education. (Bulletin No. 130, April, 1928.)

Samples of Part-Time Cooperative Schools.—The following description of part-time cooperative schools is taken from Bulletin 130 of the Federal Board—pages 10–11:

Massachusetts

"Cooperative part-time education is classified into two types—

"*a*. Part-time cooperative schools or departments distinctively organized to give instruction for a single trade in connection with one specific industrial plant. The pupils are under full control of the school and are usually under some sort of indenture to the company as well. The cycle of alternation between school and plant is commonly two weeks.

"*b*. A part-time department distinctively organized to give instruction for a single trade, and to place its pupils with various employers in that trade for their shop experience. The school has no definite agreement with any employer, and no absolute control over the pupil's work during his shop week. The cycle is commonly two weeks. Several full-time, day trade schools have become, in effect, part-time schools of this type, by placing pupils on a cooperative basis while still enrolled as school members.

"Under type *a* three schools are in operation—Beverly Industrial School, Southbridge Vocational School, and Weymouth Part-Time Cooperative School.

"The Beverly Industrial School was organized in 1909, the boys spending alternate weeks in the shop located in the plant of the United Shoe Machinery Corporation, where their shop work is taught, and in the school, where they are instructed in related drawing, related mathematics, related science and English, citizenship, history, economics, and hygiene. The shop week is commonly 44 hours, and the school week

30 hours for 50 weeks in the year. The boys spend about two years alternating between shop and school, and 9 months to a year on *full time* in shop, at the end of which they graduate. There are about 70 boys in this course (machinist), and 20 boys graduated on December 14, 1927.

"The Southbridge Vocational School was organized in 1920, with one group of boys in the factory of the American Optical Co. (machine shop) and another group in the mill of the Hamilton Woolen Co. (worsted manufacturing). The cycle in Southbridge is similar to that in Beverly, the groups alternating between the shop and the school. The opportunities for training now afforded by the American Optical Co. have been extended to other departments of their business until they now include machine-shop work, machine drafting, printing, cabinet-making, dispensing optical work, sheet-metal work, steam engineering, and carpentry. There are about 100 boys enrolled in the school.

"The Weymouth Part-Time Cooperative School was organized in 1927, the boys spending alternate weeks in the factory of the Fulton Shoe Co. and in the school. There are 14 boys in this school now.

"Under type *b* there are five schools, all located in the city of Boston, as follows: The Charlestown Part-Time Cooperative School, the Brighton Part-Time Cooperative School, the Dorchester Part-Time Cooperative School, the East Boston Part-Time Cooperative School, and the Hyde Park Part-Time Cooperative School. In each of these schools the boys spend alternate weeks in employment in industry for approximately 44 hours per week, and in school for 27½ hours per week. These schools differ from Beverly, Southbridge, and Weymouth in that a large number of concerns employ a few boys each, and the school has not the same control of the students' activities during their employment. A well-equipped shop at the school provides opportunity for instruction during short periods of unemployment.

"There are approximately 200 students in the Brighton Part-Time Cooperative School, which is organized to give instruction in automobile repair, the boys working in industry on alternate weeks. At present only the third- and fourth-year boys participate in employment outside the school, the first and second years being alternate weeks in 'school' and in the school shop. The fifth-year boys are employed full time in the industry.

"There are approximately 340 students in the Charlestown Part-Time Cooperative School, which is organized to give instruction in electrical work. The organization as to program is similar to the

Brighton Part-Time Cooperative School, except that at present the second-, third-, and fourth-year boys participate in employment outside the school, while the first-year boys alternate one week in school and one week in school shop and the fifth-year boys are employed full time in industry.

"There are approximately 200 students in the Dorchester Part-Time Cooperative School, which is organized to give instruction in carpentry, cabinetmaking, and wood pattern making. The organization of program is similar to the Brighton Part-Time Cooperative School.

"There are approximately 200 students in the East Boston Part-Time Cooperative School, which is organized to give instruction in machine-shop work. The organization as to program is similar to the Brighton Part-Time Cooperative School."

The same bulletin lists some eighty cities maintaining cooperative part-time classes. The pupil enrollment given was 5,682, and over thirty different types of occupations are given place on the various curricula.

Labor and Vocational Schools.—Modern labor and labor leaders are awake to the need for vocational training. Labor unions today invite the opportunity to participate in the organization of vocational schools. Part Three of this book includes a survey of the official utterances of the American Federation of Labor in convention upon the subject of vocational training. At this point, however, the booklet issued by the Federal Board for Vocational Education, "Labor's Responsibility in Cooperation with Employers and the Public Schools," is pertinent. William Green, president of the American Federation of Labor, states in his preface to this brochure that the American Federation has for many years recorded its approval of vocational education and vocational rehabilitation. He states also that the American Federation of Labor was primarily responsible for the passage of the Smith-Hughes Act in 1917.

Industrial Education at First Neglected the Girl.—Most of the trade or industrial training has been applicable to boys only. However, since the advent of women into industry, the problem of industrial education for women has come to the fore. Due to a variety of reasons, women have found places in industry, and hence a training for those positions has become a necessity.

Trade Schools for Girls.—The realization of the importance of the training of women for positions in the fields of textiles, millinery, and the like led to the establishing of such schools as the Manhattan Trade School for Girls in New York and the Boston Trade School for Girls.

The Manhattan School, like so many schools of this nature, was founded under private auspices in 1901. The New York City Board of Education took it over in 1910. The New York Industrial Education Survey, quoted in Mays' "The Problem of Industrial Education" (pages 318–319), states the aims:

"1. 'To train young girls, who are forced to leave school and become wage earners, to enter the skilled trades.'
"2. 'To imbue with a love and respect for work.'
"3. 'To arouse in them a desire to become the best type of workers.'"

The significance of women's place in industry is clearly shown by the adoption of private trade schools, continuation schools, and factory training schemes to meet the demand for trained women.

The Responsibility for Vocational Education.—The problem of placing the responsibility for industrial education has been a difficult one. Industrial education has, in the past, been regarded as a joint responsibility of the individual and the trade to which he seeks admission. As social theory has developed, however, it has become more and more a state function. This gradual development has been extremely slow. In fact, prior to 1862, engineering education and agricultural education were not regarded as public functions. Then the Federal authorities recognized engineering education as a function of the state and hence subject to state aid, regulation, and protection. Only since 1862, in the United States, has industrial education been maintained, in part, by public agencies.

Public Support for Vocational Education.—Arthur Mays, in "The Problem of Industrial Education," page 353, speaks of the responsibility of public agencies, and his thought can be taken as a typical present-day idea of the place of industrial education in the state. He states:

"Clearly the training in citizenship and the interpretation of the economic and social factors of industrial life and of the business of production can safely be entrusted only to agencies constituted and controlled by organized society for that purpose. Private institutions, however sincere and well meaning, scarcely can escape the temptation to color their teaching so as to serve their selfish interests. Moreover, there seems to be a further responsibility with respect to the education of industrial workers which society does not yet fully recognize, namely, to guarantee to every worker adequate training. This doubtless can be accomplished only by means of compulsory part-time education in public schools during the training period, and close, authoritative supervision of the learners' work in industry during the same period. The assumption of this responsibility under the present social conditions can, of course, be justified only in the case of minor citizens, but the duty to provide such a guarantee of vocational preparation seems to be implicit in the traditionally accepted guardianship over all minors by the state. This principle is recognized in child-labor laws, compulsory general education laws, compulsory part-time school laws, and in various other laws affecting the welfare of minor citizens. This appears to be the probable next step in the development of a democratic program of public education in the United States."

Federal Grants for Education.—Federal grants to the states in aid of public education have had an interesting history in America. Some mention has been made already of this, but since 1870 the trend toward Federal aid is marked.

Governmental aid originated in the action of the General Court of Massachusetts which, in 1659, made grants of land to certain towns for the maintenance of grammar schools. Congress prior to 1780 made use of the vast western area, ceded by the states to the central government, for the benefit of all of the states. It became customary, after 1785, for the Federal government to give such land grants to aid education. This policy definitely established the principle of Federal grants in aid of education.

The treasury surplus of 1837 was distributed among the states, and some of this was set aside by the states for use in the public schools. Later some of the money from the sale of public land was given to the states for the maintenance of their schools. In 1920 such a grant was made, when the Federal government gave to the states containing oil wells a part of the royalties and lease money paid by the oil companies. This grant must be used for highways and education. Fletcher Swift, in "Federal Aid to Public Schools" (page 19-25) notes that money grants prior to 1920 were so mismanaged by the states as to be an insignificant factor in the support of public education.

Until the passage of the Morrill Act (or Land-Grant College Act, already discussed) no specific requirements were made by the Federal government of the states. The interesting and epoch-making Morrill Act has already been treated in this volume. It is sufficient to note here that the Morrill Act was amended in 1890, and again in 1907. These changes, however, made no essential modifications in the original act.

Recent Legislation Concerning Vocational Education and Rehabilitation.—The Hatch Act, passed in 1887, provided funds for the establishment of experiment stations at agricultural and mechanical colleges. In 1906 this act was amended so that the funds could be increased. Although little governmental control of funds is permitted by this act, very definite prescription is made as to the uses to which the money can be put. The basic idea seemed to be to have the Federal government aid state projects in education which might prove to be of national or universal benefit.

The year 1914 saw a very different type of Federal aid, an aid based upon a different philosophy. This was the offer made by the central government to meet dollar for dollar money expended by the states for certain types of educational work. It might be said that the policy of the central government expressed here is one of stimulation rather than mere aid. The acts which

SUMMARY

express this idea include the Smith-Lever Act of 1914, the Smith-Hughes Act of 1917, the Fess-Kenyon Act of 1920, and the Sheppard-Towner Act of 1921.

The Smith-Lever Act provided money to be distributed among the states on the dollar for dollar basis, for the purpose of developing and maintaining extension work in home economics and agriculture.

The Smith-Hughes Act provides for the similar distribution of funds for the encouragement of vocational education of less than college grade.

The Fess-Kenyon Act provides for the vocational rehabilitation of persons injured in industry or otherwise incapacitated for a normal place in industry.

The Sheppard-Towner Act provides for the promotion of the welfare and hygiene of maternity and infancy.

These acts, the importance of which is tremendous, are more fully described in next section of this work.

All those laws show in their clauses an increase in the definiteness with which the uses of the grants are defined, and each provides for real governmental supervision. Arthur Mays says that the Smith-Hughes Act "seems to be the high point in all Federal laws appropriating funds to be distributed among the states." ("Problem of Industrial Education," page 360.) It is, no doubt, the chief expression of the present-day policy of the central government with reference to grants in aid of education conducted by the states.

The question as to the merits of Federal aid is beyond the scope of this work. The desirability of Federal aid or Federal authority in questions of education is a debatable matter. It appears, however, that growing Federal authority is a normal expression of the modern trends in economic, social, and political life.

Summary.—The Federal Board for Vocational Education is responsible for the administration of six separate acts of Congress dealing with vocational education and vocational rehabilitation:

1. The Smith-Hughes Act (1917) for the promotion of vocational education.
2. The Fess-Kenyon Act (1920) which provides for the promotion of vocational rehabilitation of persons disabled in industry or otherwise and their return to civil employment. (This Act was first passed in 1920, extended in 1924, and again in 1930.)
3. The act extending the benefits of the Smith-Hughes Act to the Territory of Hawaii (1924).
4. The George-Reed Act (1929) which provides for the further development of vocational education in the several states and territories.
5. The act providing for the vocational rehabilitation of disabled residents of the District of Columbia (1929).

Industrial Schools for Poor or Delinquent Children.—One of the first applications of the results of the experiments of Pestalozzi and Fellenberg was to the industrial schools for (a) paupers, orphans, and deserted children; and (b) children and youth who had been guilty of some criminal offense and been imprisoned for corrective purposes. In the early days of these widely divergent types of school, seldom were they separated institutions. Later, however, a realization of the injustice of non-classification brought about the divorce of one from the other. Many of the stories of Charles Dickens are source material for a picture of the times. The first type of school became known as a hospital or orphan home, while the second type became known as a farm school, house of refuge, or reform school. In England it is important to note the presence of a third type, a school of industry. This was attached to the elementary school for the manual labor class of society.

German Schools for Destitute Children.—John Falk (1770–1826) established an industrial school at Weimar, Germany, in 1820. Educated at Halle, Falk at first tried his hand at poetry. He noted the great numbers of boys running wild over the German battlefields. He determined to do something to assist these scavengers, who had been brought from all parts of the country to Jena and Leipsic and the battle countries. To secure funds to deal with these destitutes Falk begged for money. In 1818 he organized the "Friends in Need." In 1820 he had 300 children in his own home. He wrote to a friend in 1821: "The children of robbers and murderers sing psalms and pray; boys are making locks out of the insulting iron which was destined for their hands and feet, and are building houses which they formerly delighted to break open." (Quoted from Bennett—"History of Manual and Industrial Education up to 1870," page 211.) The boys, with the aid of journeymen, built the houses in which they lived, and equipped them with the necessary requisites for home life.

Baron Kottwitz, at about the same time, carried through a similar project in Berlin for the factory children.

Frederick William's Activities in the Field of Education.—In 1824 Frederick William I of Prussia established an institution at Potsdam for the maintenance and education of the orphans of soldiers. The desire to utilize the boys as recruits for the army and the presence of a fertile field for training them soldiers, prompted Frederick William in this venture, although it is said that he modeled his school on Francke's Institute at Halle. The orphans were required to pay for their education by serving twelve years in the Prussian army, three of which might be spent in a school for non-commissioned officers. There were attempts to introduce industrial pursuits at this school, all of which failed.

It was found unprofitable for the orphans to try the manufacture of Brabant lace and silk.

Bache's Visit to Europe, 1837.—Alexander Bache (1806–1867) visited Europe in the hope of finding material to incorporate in Girard College, Philadelphia. This grandson of Benjamin Franklin visited the Potsdam orphan school in 1837. There he found three educational divisions, the elementary school, the trade school, and the music school. The aim of the trade school, Bache says in the "Report of Education in Europe" (Girard College, Philadelphia, 1839, page 123), was "to economize the funds of the institution, by making within its walls articles of clothing required for the pupils, but more to secure the acquisition, not only of general mechanical dexterity, but of a trade, which may serve to increase their emoluments when they enter the military service."

The trades taught in 1837 were those of blacksmith, saddler, tailor, shoemaker, and lithographer.

Orphan Schools in England.—England, too, felt the orphan school movement. In 1685 William Blake of Covent Garden founded the "Ladies Charity School" or the "Hospital at Highgate." He purchased a house and expended £5,000 in this benevolent project. Blake is usually credited with being the originator of the charity school movement.

A similar school was founded about the same time at Kensington. In both schools the fatherless and the poor were clothed, fed, instructed, and put out to work at various trades. There is no evidence that the trades were taught in the schools, and it is to be inferred that the orphans assisted with the household upkeep only.

Bache notes that Great Britain exceeded all other European countries in the number of charity schools at the time of his visit. He graphically describes several, and, especially, deals with Heriot's Hospital in the city of Edinburgh. In 1623, Heriot, a jeweler of Edinburgh, provided in his will for the maintenance and education of poor, fatherless children of Edinburgh. Between 1630 and 1650, a large Gothic building was erected for the institution. Boys were bound out as apprentices and received payment for their services, while supervised, at the same time, by the institution. There was instruction in drawing, the only manual arts education given. The methods and course of study closely resembled those of the Prussian schools of the time.

The famous Christ's Hospital, or Blue-Coat School, in London, was an orphan school of this type incorporated by Edward VI in 1553.

The graduates of these schools followed one of several paths; they might become apprenticed to a master tradesman or to a merchant, or they might enter the navy or merchant marine. A few also went to the universities.

The Industrial School as a Reformatory.—Soon the idea spread that the industrial school was an admirable reformatory for the juvenile criminal. The notion of aiding the delinquent by education as opposed to the notion of leaving him to his life of crime was a great step in social history. Francke was perhaps the first to suggest the segregation of the juvenile criminal from the adult criminal.

In line with these ideas the Redemption Institute at Horn, near Hamburg, Germany, was founded by an association of socially minded people in 1833. The purpose was the reclaiming of the abandoned children of the lower classes. T. H. Wichern, who conducted the institution, received the boys into his own house. Says Henry Barnard in his "National Education in Europe," page 517:

"His [Wichern's] first step was to procure a plain dwelling, and to remove everything from within and without which gave it the appearance of a place of punishment or correction." Again—"By forgetting or forgiving the past, and encouraging every effort on the part of these depraved outcasts of society . . . Mr. Wichern succeeded in working remarkable changes in the character of a large majority of all who became inmates of his family." (From Bennett's Source Material, page 217.)

The work, for these "outcasts" was farm work and the usual household duties. As the institute grew, a bakery, workshop, stable, library, chapel, and school house were added. The establishment was divided later into three parts: (1) a reform school for 65 boys and 35 girls; (2) an institute of brothers, a normal school for the training of individuals to carry on similar enterprises elsewhere; (3) a printing house and bookstore. Barnard notes the activities of the boys and lists gardening, farming, shoemaking, carpentry, baking, spinning, printing, lithographing, wood engraving, and bookbinding, among them. The girls had household duties and some out-of-door work.

Griscom's Year in Europe (1818).—John Griscom in his "A Year in Europe" visited during 1818 an English institution conducted by the Philanthropic Society. As early as 1788 this society provided for an institution to be a means of industrial, moral, and intellectual instruction for juvenile criminals and the destitute offspring of convicted felons.

The first fruit of this was a hired house near London at St. George's Fields, where some dozen children were collected and placed in charge of a man and his wife who guided the "destitutes" through manual labor.

Soon the plant came to embrace three houses, one devoted to shoemaking, a second to tailoring, and the other to carpentry. Griscom (page 171) notes:

"The boys receive a sufficient share of school learning, and are placed, on their admission, in one of the various manufactories or workshops, which are conducted by master workmen and journeymen. The principal trades pursued are printing, copper-plate printing, bookbinding, shoemaking, tailoring, rope-making, and twine spinning. A portion of each boy's earnings goes to his credit, and is given to him at his discharge.

"Besides receiving those poor juvenile offenders in their establishments, the committee have adopted the plan of apprenticing out some of the best-behaved boys to tradesmen of good character, with a sufficient premium; but they are still considered as under the care of the society. The girls make their own clothing, and shirts for the boys; wash and mend for the manufactory; and, in short, are educated so as to qualify them for useful and respectable service. About one-hundred and fifty boys are within the walls and more than fifty girls." (Bennett, page 219.)

Sidney Turner became manager of this institution in 1846, and later he and William E. Gladstone, the treasurer of the Philanthropic Society, visited France to survey the industrial schools there. When they returned to England, they worked out a plan for a reform school wherein farm labor should be the principal occupation, seconded by trade work and the handicrafts. A site of 140 acres was purchased, Red Hill Farm in Surrey.

The school opened with three boys in 1849. By the end of the first year the enrollment had mounted to 65. The school continued to grow, and by 1871, the plant consisted of 300 acres with several houses, each providing for about 60 boys.

George C. T. Bartley, a Londoner, in "The Schools of the People" published 1871 (pages 261-262) notes that the schoolroom activities of the boys occupied only three hours on alternate days in summer and a little more in winter. The remainder of the time was spent in industrial occupations. About two-thirds of the boys were occupied in field work and one-third in the various industries of the institution which included house work, laundry work, baking, tailoring, shoemaking, carpentry, blacksmithing, etc.

The Ragged Schools of Great Britain.—Before 1870 there was no system of compulsory education in England. The effect of the Industrial Revolution on industry has been noted, and now its effect on education may be observed. The Education Act of 1870 provided for a local school tax, a representative local school authority, and compulsory attendance of children.

Before this Act, the education of children of the poor depended upon private philanthropy and public grants. Philanthropy was widespread, but despite the generosity of the few, many children grew up without schooling. Dr. Thomas Guthrie (1803-1807), a leading Scottish educator, noted that the child's passage to school was through the police office, and the passport was a conviction. In his "Seed Time and Harvest of Ragged Schools" (page 11), Edinburgh, 1860, he pointed out that the juvenile delinquent schools were about the only schools wherein the poor might learn. He also noted that even if there were schools for the poor, many could not attend because they had to engage in gainful occupations in order to keep body and spirit together.

"Christian missionary zeal was the backbone of the 'Ragged School' movement," says C. A. Bennett on page 223 of his "History of Manual and Industrial Arts up to 1870," and the Bible was the chief textbook.

Labor Unrest Leads to Education.—Fear of revolution also can be added, for during the early years of the reign of Victoria (1843–1848), due to economic pressure, there was general labor unrest. Crime was rampant and poverty widespread, and conditions grew worse rather than better.

Early Ragged Schools.—Perhaps the earliest of the Ragged Schools was that of John Pounds, a shoemaker of Portsmouth, established in 1819. In his small and wretched shop he often collected the ragged children of Portsmouth to give them lessons of honesty, sobriety, and truthfulness. He also taught them to read, and aided them in finding means to make honest livings. In order to get children to attend he baited them with food, a tempting thing for hungry boys. It is said that some 500 children were members of his cobblers' class during the twenty years that he taught while he worked. But this was merely a beginning, inefficient and crude.

A more effective start was that instituted by the sheriff of Aberdeen in 1841. Following this, nearly all the cities of England and Scotland began to have similar schools. The name "ragged" was applied in 1842 in a public appeal for funds by a new school. Lord Ashley noted this, and backed up the movement, which enlisted the support of many of the leaders in society including the Queen herself. In 1844 The Ragged School Union was formed, with Ashley as president. (Bennett, page 224.)

The Ragged Schools attempted to meet a very definite need created by a mal-adjusted industrial system. They were a social necessity which served to ameliorate a condition growing out of an economic difficulty. The teachers in these schools were volunteer workers who served without financial recompense. All classes joined in the work that was felt to be so necessary.

The schools themselves were certainly individualistic. There was no uniformity of curriculum, no standards, no promotions. Each school taught that which it could. Instruction in the Bible and the elements of religion predominated. Industrial work was considered an important factor in respectable living, yet no provision for this work obtained in many of these schools.

There were day schools, night schools, and Sunday schools of this type.

From Bennett's history, page 225, is this typical winter day at Dr. Guthrie's, Edinburgh:

```
 8.00- 8.30 Ablutions
 8.30- 9.30 All working
 9.30-10.15 Breakfast and play
10.15-11.00 Calling roll and Bible lesson
11.00- 1.00 One half in schoolroom and other half in work room
 1.00- 2.00 All walking
 2.00- 2.30 Dinner
 3.00- 5.00 One half in schoolroom and other half in work room
 5.00- 6.30 All working
 6.30- 7.15 Supper and closing
```

The industrial work, when given, was largely that which the teachers could teach, and also that which was necessary in order to give the youth ability to gain employment.

In 1855 London had fifty Ragged Schools with industrial classes, and these included some 2,000 children.

The "Transactions of the National Association for the Promotion of Social Science, 1859" (page 449) indicates a reduction in juvenile crime, and lays this to the influence of the Ragged Schools in a large measure. Other reports agree that the fall in juvenile crime can be credited to the work of the schools. Perhaps most striking is a statement in C. F. Montagne's "Sixty Years in Waifdom, or The Ragged School Movement in English History" (page 290) that during 1864 the London Ragged Schools placed 1,920 domestic servants.

At least, the Ragged Schools prepared the way for free compulsory elementary education at public expense by educating the public and by showing that the fruits of education are tangible and very real. They were also the forerunners of the technical schools of London, into which many of them developed.

The First Grant by Parliament for Elementary Education, 1833.—Even up to 1861, when the revised code governing elementary education in England was adopted, the children of the poor were educated by public grants and private bequests. The rich and the middle classes made education a private matter, supported by tuition fees and private gifts and endowments. In 1833 the first public grant to elementary education was made by the government, and the amount of £20,000 was used only for the erection of school buildings. There was no attempt before 1860 to provide free public elementary education for all classes.

Condition of the Industrial Schools in the Nineteenth Century.—The industrial schools of the period were perhaps the poorest of all of the schools in that the parents of the students were unsympathetic to the movement which withdrew their children from work in the factories or fields. To meet this objection it has been shown how the schools incorporated work among their activities. Until 1843 Reform Schools and Ragged Schools were considered as about the same.

In 1843 the Committee of the Council on Education asked for a report on the success which had attended the operation of industrial schools. Upon this report conditions were set forth under which annual grants could be expected.

In 1854 "Certified Reformatory Schools" were established, which gave lodging and board to children. In 1857 an Act was passed which encouraged the transformation of refuge Ragged schools into Certified Industrial Schools. Children under fifteen, upon sentence of vagrancy, might be sent by a justice to these schools. Thus the Ragged Schools became in part supported by the government.

Continental education, and especially industrial education in the elementary schools, differed greatly in its evolution from English education of the same type. The origin of both systems was the same; the work of Pestalozzi, Fellenberg, and Wehrli constituted the cornerstone. However, compulsory education came earlier in Germany than in England. In Weimar, compulsory education for all children between the ages of 6 to 12 obtained in 1619. In England compulsory education did not come until 1870. All through the German states compulsory education had been established, and after the revolution of 1848 most of Germany had free elementary education.

In 1839 the German law required that a child must receive three years of regular school instruction before he could procure a working certificate to allow him to be employed in a factory before he was 16 years old. Germany removed the economic barrier which held the English child from school. The German schools developed drawing and needlework rather than industrial arts in the schools during this period.

INDUSTRIAL EDUCATION IN AMERICA

Education Considered a State Function in America.—America accepted the principle of education as a state function and at public expense, but during Colonial times the very poor were educated, if at all, by private societies of a philanthropic or religious nature. Later institutions received aid from the government, in addition to aid from other sources, for the education of the orphans, the physically handicapped, the Negroes, the Indians, etc.

Early Schools.—One of the first of the institutions to deal with these children had the industrial idea deeply rooted in its soul. It was the Farm and Trades School of Boston, established privately in 1814. In 1833 it was removed to Thompson's Island (157 acres) in Boston Harbor, where it still continues to flourish. Nearly a dozen trades and occupations are taught in this school. Orphan asylums in America embracing industrial courses grew in great numbers during the nineteenth century. Girard College, Philadelphia, is an example of this type.

Kiddle and Schem in "The Cyclopedia of Education," page 466, note that at the age of 12 years the Girard College boy began to receive practical experience along shop lines. In 1864 provision was made for instruction in type-setting, printing, bookbinding, type-founding, stereotyping, turning, carpentry, photographing, electrotyping, electroplating, and telegraphy.

Industrial Schools for the Freed Negro.—At the close of the Civil War the freed Negro presented a problem in employment and vocational training. To aid in the adjustments the Freedman's Bureau was established. Education for Negroes became a function of this bureau. General Armstrong was one of the bureau's agents, and was a capable man to assist in this work. His early life had been spent in Hawaii. He knew the American Negro as few men in the North knew him. He realized that many Negroes could be trained to lead their race. He suggested the purchase of a tract of land on the Hampton River to the American Missionary Association. The society followed this advice, and established the famous Hampton Normal and Industrial Institute with Armstrong as principal. The school opened in 1868. Gradually more and more trades were added to the work at Hampton until by 1884 a machine shop paved the way for Hampton's becoming one of the model trade schools of the country and a pattern upon which many schools have been devised, especially for the Negro and Indian.

The Reform School in America.—America, also, fell in with the industrial reform school idea. One of the first to be established was the New York House of Refuge on Randall's Island (1824) for reformation of juvenile delinquents. In 1828 Philadelphia had its House of Refuge, while Boston had established the House of Reformation for Juvenile Offenders.

Snedden, in page 115 of his "Administration and Educational Work of Juvenile Reform Schools," quotes the aim of these institutions as given in an opinion of the Supreme Court of Pennsylvania handed down in a case involving the Philadelphia House of Refuge. It was "reformation, by training its inmates to industry, imbuing their minds with the principles of morality and religion, by furnishing them with the means to earn a living, and, above all, by separating them from the corrupting influence of improper associates."

The prison labor at that time was organized in a similar fashion to that of the reformatories, and the Prison Discipline Society stated: "in prisons where labor is prevalent there is no gambling, profane swearing, Sabbath breaking, but vast moral training." The report mentioned Sing Sing and Auburn and the New York House of Refuge as exemplary institutions. (Snedden, page 116.)

The early industrial "workhouses" taught occupations which were hardly considered trades, and Bennett ("History of Manual and Industrial Education") remarks, "it was not the trade he [the inmate] was likely to follow as a means of livelihood after leaving the institutions." (Page 248.)

Many reform schools were established after 1830, and if little in the way of industrial education characterized them it is because, for the most part, the inmates were below 14 years of age.

The Schools Supply the Need for Industrial Education Caused by the Decay in the Apprentice System.—From the economic point of view it has been noted how the Industrial Revolution put an end, in large part, to the apprenticeship system that had perfected itself through the medieval guilds. The factory system, with its specialization of function, its mass production, and its impersonal relationships left no place for the apprentice training. Schools, either factory or otherwise, were looked to, to give aid to young men entering industry who formerly could have been apprenticed to journeymen or masters. The trade processes had to be taught, technical knowledge related to the trade had to be imparted, and general schooling and moral discipline also had to form a part of the youth's training.

Apprenticeship the Basis of Trade Education.—Apprenticeship was the basic form of European and American trade education until after the factory system had been established. Where the effects of the Industrial Revolution were slow in taking root, the apprentice system persisted. In English-speaking countries the decline of the apprenticeship system was most marked during the last quarter of the eighteenth century and the first half of the nineteenth century. The textile industries making use of the labor-saving devices and great inventions of the Industrial Revolution led the attack on the time-hallowed medieval methods. The inventions and the progress of the Industrial Revolution have been noted. It remains to discuss the educational effects.

Sunday schools, part-time schools, factory schools, and continuation schools for factory workers arose to meet the needs of a new industrial, economic and social structure.

Parliament Passes the Factory Act of 1802.—Sir Robert Peel was a champion of the Factory Act of 1802. Peel was a manufacturer himself, and his interest in the apprentices is a noteworthy start in a changed attitude on the part of business men toward labor. The Act was an advance in that it provided for the clothing, feeding, and instruction of apprentices. Night work was abolished for them and the working day was reduced to twelve hours a day, excluding mealtimes. Another provision of that epoch-making act was for factory inspection.

The rapid development of the Industrial Revolution and the resulting urbanization of industry played havoc with the Act of 1802. Employers found child labor plentiful in the towns, and apprenticeship agreements failed to be observed.

The Factory Act of 1819.—Again Sir Robert Peel came to the aid of the apprentice, and the Act of 1819 was passed after a four-year struggle. This provided that "no child under nine years of age should be allowed to work in a cotton factory, and no young person under sixteen to work more than twelve hours per day." By 1825 the age limit had been raised from sixteen to eighteen and the number of hours per week was reduced from seventy-two to sixty-nine. This Act dealt only with the cotton industry, and hence failed as a general remedy. It did not provide schooling for the children.

The Factory Act of 1833.—The Act of 1833 was the first to provide for schooling. The horrors of the factory system featured English history during the decade of the twenties, and by the end of this period public interest and public action were aroused. The influence of Charles Dickens, who, through his novels, portrayed the conditions of the children in words of plain and naked truth, is well known. The influence of such men as Lord Ashley (Anthony Cooper) and Robert Southey among many others is also worthy of note. The pictures painted by Dickens were sufficiently vivid to arouse the public to action.

Ashley saw a remedy in a ten-hour law. He failed to get an act providing for this through Parliament, but his fight for it resulted in the passage of the Factory Act of 1833. Graham Balfour in his "Educational Systems of Great Britain and Ireland" (page 47) says, "In respect to education [the act] was immensely in advance of all provision for the working class at that time." Under this statute children between nine and thirteen might be employed only if they had a voucher of having attended school two hours on six days in each preceding week. The inspector might require the employer to make a deduction of one penny in the shilling from a child's wages, and pay the same for the schooling of the child according to his direction.

Here, perhaps for the first time, the close relationship of labor and education can be noted. This law was amended time and again and finally a new law, the Act of 1844, was passed making the hours of schooling longer. It is evident, however, that a change had come in the attitude of the public toward the function of the government in industry and in education.

Further Legislation of a Similar Nature.—Thus there came into being the half-time schools. At first these were operated only in the textile industry. Later acts, especially that of 1874, extended the field to include the non-textile factories.

Age requirements also were raised in succeeding laws. The law of 1874 raised the minimum age from eight to ten, the Act of 1893, from ten to eleven, and that of 1901, from eleven to twelve.

The half-time school has been a point of contention since its inception. Early in the twentieth century public opinion favored its abolition. This saw fruit in the Education Act of 1918 which provided for the closing of the half-time school. No date has as yet (1930) been set for the enforcement of this clause of the act.

Hence in England the part-time school was offered as a substitute for the apprenticeship system.

The Decay of the Apprenticeship System in France.—In France the guild system died a hard death. The Revolution acting through the constituent Assembly definitely abolished the guilds in 1791. Under the provisions of this act, each individual could select for himself the occupation which he desired to pursue, and could enter upon work therein when he obtained a license and conformed to certain trade regulations. The power of the guilds, then, was assumed by the government. As no substitute for apprenticeship was devised, the plan worked badly.

The French law of 1851 planned to regulate apprenticeship by defining the responsibilities of the master as to instruction. France early turned to trade and technical schooling as a substitute for apprenticeship. France had highly developed engineering schools at the time, and in them it is not difficult to find a root for the development of trade training.

In 1788 the Duc de La Rochefoucauld established a school on his estate at La Montagne. This school was for the purpose of instructing the sons of the non-commissioned officers of the Duke's regiment in general education and in the elements of a trade. In 1799 the government became so impressed with the work of the institution that it declared it a national school, and removed it to Compiègne. Bonaparte, who visited it a few years later, declared that he found there "workmen distinguished in their craft, having great dexterity in execution, but hardly one who can make a drawing of the simplest type of machine or could express idea by a sketch. It is a gap in French industry; I will fill it up. No more Latin, but trades with the theory necessary for progress. Here excellent foremen will be trained" ("L'Enseignement Technique en France," Tome 1, page 271.) In 1806 the school was further enlarged and moved to the plant of a former convent at Châlons-sur-Marne. This was the beginning of the French series of Ecoles Nationales d'Arts et Métiers.

It is well to pause to survey this school at Châlons. "Boys were admitted at eight years of age. Up to twelve years they were taught reading, writing, the elements of French grammar, arithmetic, the elements of geometry, and drawing. Later came the study of descriptive geometry applied to crafts, drawing and tinting applied to drawings of machines, and the principles of mechanics." (Bennett—"History of Manual and Industrial Education," page 277.) The shops included those of the blacksmith, machinist, metal turner, foundryman, carpenter, cabinet maker, wood turner, and wheelwright.

After 1826 the entrance age was raised to fifteen years. The course was made four years—part of the time in the shop, the rest in the academic work. Again in 1832 the entrance age was placed between fifteen and seventeen years.

Many schools of this type followed. In 1815 one was established at Beaupreau, a third opened at Aix in 1843. These schools all had for their aim the production of superior workmen who knew the related mathematics, science, and other academic knowledge of their trades to enable them to fill positions as foremen, superintendents, and technical experts.

France Takes the Lead in Technical Education.—A national commission on technical education was established in France during 1863. Its report was published in 1865. One of the fields in which this commission took an especial interest was that of child labor in the factories. Proceeding on the thesis that the law of 1841 in regard to child labor was inadequate, they urged the raising of the admission age of children to industry to ten years, and the reduction of the child's working day.

This quotation from "The French Commission on Technical Instruction" indicates the scope of the inquiry:

"The possibility of thus rendering the labor of children in manufactories compatible with the instruction they need is proved in that part of England by the establishment of half-time schools in which children of from 8–13 years of age must not, according to law, be occupied more than six hours a day. In England, as well as in Belgium, it has been remarked that children who for half the day are subject to the discipline of the workshop, pay great attention to their lessons at school, and acquire the instruction there given about as rapidly as those who go to school the whole day."

The recommendations were: (1) The day of the 10–13-years-old child should be six hours; (2) a portion of the remainder of the day should be spent in school. It was noted that the enforcement of this recommendation might be simplified if the children attended school run by the employer; (3) for the 13–16-year group, the commission suggested a ten-hour working day. A two-hour evening period, the commission thought, could be spent at classes and lessons. ("The French

Commission on Technical Instruction"—translated and printed by Eyre and Spottiswoode, London, 1868.)

This memorable commission recognized a diversity of purpose in shop instruction. It indicated a group whose purpose was merely the training of apprentices, i.e. the orphan school group and the agriculture group. It noted a second group where a trade was taught in addition to regular secondary education. This type is exemplified by the workshops of the Christian Brothers. And it also recognized an aim in training for industrial leadership by the teaching of the theory, principles, and practices of industries. The commission remarked that these aims were wholesome and worthy of state aid and encouragement.

In the record of the hearings of the French Commission it is interesting to note the view of M. Bader, the director of the Technical School of Mulhause (page 94 of the "Report of the French Commission on Technical Instruction"). He states that "pupils work in the carpenters' shop for these one-hour periods during the week. They are not expected to become expert mechanics at the end of the term, but rather they are taught to form an opinion of manual labor, and a coordination of hand and eye. The director considers the work as similar to an athletic exercise." This idea is of interest in view of the American attitude toward the manual training movement of the later nineties.

German Factory Legislation.—In 1853 the Prussian government prohibited the employment of children under twelve years of age in factories. At the same time, as noted heretofore, a Prussian guardian of a child under sixteen years had to certify that the child could read and write if he wished to place him in industry. Children under fourteen could work for six hours a day, provided that three hours per day were spent in instruction.

The Sunday School of Germany a Root of the Continuation School.—As Germany led the world in compulsory school attendance laws, so also she developed a system of compulsory continuation schools. One of the roots of the continuation school was the Sunday school of earlier times. In Wurtemberg and

Bavaria by 1803 attendance was compulsory at the Sunday school. M. E. Sadler in his "Continuation Schools in England and Elsewhere," page 520, notes that a man could not marry unless he first produced a certificate that he had gone through a course at a Sunday school. The Sunday school taught writing and reading as well as religion.

With the coming of a new industrial system, a new use was found for the Sunday school. England and France were surpassing Germany in the development of modern industry, and the German employers began to realize their need for trained workers to meet the competition. The continuation school rose to fill this demand.

After much inadequate legislation, the Act for Regulation of Industry in 1869 provided a decisive step in the enforcement of attendance at continuation schools. Employers were "compelled to allow their workmen under 18 years of age to attend a recognized continuation school, and the communes were empowered to frame by-laws making attendance at such schools obligatory on all workmen under eighteen." (Quoted by M. E. Sadler, "Continuation Schools in England and Elsewhere," page 522.) This was the origin of what has been a gigantic development which has attracted the interest of educators the world over.

In Germany the Industrial Revolution was more of an evolution than revolution. There was no rapid break-down of the old system, and the apprenticeship system held its position as the basic method of trade and industrial education. The complete trade school was slow in its advent in Germany, while the demand for continuation schools was high. As the workers were learning their trades on the job in the shop of the master, the half-time schools provided less shopwork training and more relatable material.

Bache notes a school wherein this was not the rule. This school might be said to contain the germ of the modern co-operative school. It was the school at Berlin, called the Institute of Trades, supported, in part, by Baron von Seydlitz. In this school, theoretical studies were given in the winter, while the summer was devoted to practical instruction.

Lyon Playfair, who wrote "Industrial Instruction on the Continent," published in London, 1852, is quoted in Bennett ("History of Manual and Industrial Education," page 291) as saying that the Berlin school had to give up the practical shopwork instructing "as might have been anticipated." Such instruction "was found to be of little advantage." Further he states: "shop instruction has been abandoned by almost all the schools; . . . in addition of the folly of attempting to teach the practice of an art within the confines of an institution chiefly devoted to other objects, it was found highly detrimental to the progress of the students, who were glad to escape from the mental labor of the classes to the muscular labor of the workshop."

Bennett comments, page 291, "Lord Playfair was a partisan in a controversy that continued for many years." The right basis for settlement of it did not come until about twenty years later, when it was found that the same principles of pedagogical analysis must be applied to teaching shopwork as had been applied in teaching other school subjects.

The Mechanics' Institute Movement.—One of the characteristics of the early nineteenth century was the demand for more education by the common people in Great Britain and the United States. The great increase in school population is a witness to this fact. A final result, perhaps, is the great twentieth century demand for college and post-college education.

One of the first non-ecclesiastical adult schools was that at Nottingham, which opened in 1798. This was an outgrowth of the Sunday-school idea, which was, in turn, the outgrowth of the church catechism class. An impetus to the movement was furnished by the numerous inventions in applied physical science and the formation of "Literary and Philosophical Societies" for mutual education.

To illustrate the humble origin of many an educational device it is interesting to note the contribution of Dr. George Birkbeck (1776–1841), who was an instructor in the Andersonian Institute at Glasgow. Dr. Birkbeck was so impressed with the ignorance of the theory of mechanics of his workmen who assisted him in his construction of models that he planned a system of lectures in scientific knowledge for persons engaged in the mechanical arts.

To the first of these lectures he attracted seventy-five persons, to the third, three hundred; and four hundred workmen attended the fourth (figures given by Fabian Ware, "Educational Foundations of Trade and Industry," page 21).

This series of lectures is noted by many educators as the beginning of the Mechanics' Institute Movement. The year was 1800.

The mechanics of Glasgow established for themselves the Glasgow Mechanics Institute in 1823. This followed the establishment of the Edinburgh School of Arts in 1821, whose aim was instruction for the laboring classes. In this institution a two year course embraced chemistry, mechanical philosophy, mechanics, geometry, arithmetic, architecture, and drawing.

In 1823 London saw the first publication devoted exclusively to interests of the working classes. This was the weekly "Mechanics 'Magazine'." J. C. Robertson and Thomas Hodgskin, the editors, devoted an early number to a survey of the work done by Birkbeck at Glasgow. Birkbeck, himself, agreed to assist in the formation of an Institute in London. On November 11 a meeting was held, and Birkbeck was the principal speaker.

Burns in his "History of Birkbeck College," page 27, notes the following: "No Tories appeared at the meeting, certain that such an establishment would result in the destruction of the empire; others remained away, fearing that the teaching of mathematics would lead to skepticism." (Quoted from Bennett, "History of Manual and Industrial Education," page 302.)

"A contribution of twenty pounds from Lord Brougham started the new institution. Birkbeck contributed his years of experience, and drew up the rules and orders." In these he sums up his aims: "The object proposed to be obtained is the instruction of the members in the principles of the arts they practice, and in the various branches of science and useful

THE MECHANICS' INSTITUTE MOVEMENT 149

knowledge." ("Rules and Orders of the London Mechanics Institution," page 5, as quoted in Bennett, "History of Manual and Industrial Education," page 303.)

The institution opened in February 1824, under the direction of six officers and a committee of thirty, two-thirds of whom were members of the working classes. The early years of the institution saw it serving about 1,000 members, who paid a small fee as tuition. The courses included physical science, mechanics, and mathematics, especially in relation to perspective, architecture, mensuration, and navigation.

The idea of this institution spread, and during the next two decades many such institutions were established. By 1841 Britain had 216 such schools. Then a notable falling off in the enrollments occurred. Bennett ("History of Manual and Industrial Education") gives a twofold explanation of this, (1) the failure of the institutes to procure competent lecturers, and (2) the failure on the part of the learners, untrained as they were, to understand the lectures. The lack of elementary education on the part of the learners precluded the possibility of their learning the more advanced subjects of science and mathematics.

A third difficulty might be given. It lies in the fact that although the schools were founded primarily for the mechanic class, the members were recruited, for the most part, from the commercial group. J. W. Hudson, in the "History of Adult Education," pages 131-132, gives a seven-year summary of membership at the Manchester Mechanics' Institute (1835-1841). It shows that of 1,184 members 328 were professional men, including artists, architects, schoolmasters, etc., 374 clerks, and only 309 mechanics. (Bennett, page 307.)

The institutions relied largely upon the patronage of the wealthy. This fact afforded another cause for the decay of this type of school.

The net result of the movement appears to be failure. The institutes either died out or lingered on as men's clubs. The contribution of the whole movement was the striking way in which it showed the need of the nation for a comprehensive system of training available for every child. Many of the institutes also developed into high-grade technical colleges—for example, the London Mechanics' Institute has become Birkbeck College.

It is interesting to note that the mechanics' institutes did not develop to any great extent on the Continent. The comprehensive elementary education, which was largely compulsory, did not necessitate the aid of institutes.

The Mechanics' Institutes in the United States.—The United States had a "mechanics' institute" movement as well as Great Britain. The root of the American movement was the same as that of the British—the effort on the part of the industrial workers to better their social and economic status through education, and the desire of the employers for more efficient workers.

The fact that American industry was more fluid and varied in type caused the institutes to be widely different and less fixed in character.

In "The One Hundred and Twenty-fifth Annual Report of the General Society of Mechanics and Tradesmen of the City of New York (1911)" it is noted that the society was founded in 1785 by mechanics, and that its objects were "mutual assistance in case of sickness or distress, and care for the widows and orphans of mechanics who should die without property." This society in 1820 opened a library for apprentices and established a mechanics' school.

Another such library was opened in Boston in 1820 by William Wood. Wood came to New York to assist in the establishment of an apprentice library for children who wished to become readers. It is to be remembered that the New York public school system was not established until 1853, and charity schools were the only places where the children of the poorer classes could receive aid. The Mechanics and Tradesmen's School was popular, and was well attended. It closed as a day school in 1858 but it continued as a night school. Its present (1929) service is free instruction for apprentices and journeymen in architectural and mechanical drawing. Its location is West Forty-fourth Street, New York City.

A second, and equally famous, institute was founded in 1824 by Samuel Vaughan Merrick and William Keating. This was the Franklin Institute in Philadelphia. The curriculum as given in Bennett's "History of Manual and Industrial Education" indicates merely that the institute attempted to broaden secondary education. The divisions of the curriculum were English classics, mathematics and sciences, and modern languages. The school remained in operation until 1832. Perhaps the Central High School, founded in 1838, is the posthumous offspring of the old institute.

Boston and Cincinnati also added institutes to the various schools in their respective communities. But, even as in England, the progress was often nil due to the insufficient preparation in fundamentals on the part of the learners.

The Lyceums in America.—As the institutes were founded in the few large cities of the country, Josiah Holbrook in 1826 published a plan to popularize education of a practical nature in the small centers. This was the lyceum idea. Town lyceums were to be formed by farmers for mutual education. The principles of chemistry of soil and scientific farming were to be taught at lyceums in the hamlets and on the farms. The New England states established many such centers, and their influence is not to be neglected.

The lyceum's contribution was just another step in the popularizing of "useful education" in the United States. The lyceums subsided after some ten years of development.

Higher Technical Education.—Technical education is a response to demands for experts in the fields of engineering, science, and allied lines. New discoveries or inventions bring with them demands which must be met by educational devices. Sometimes the demand is for research experts of great knowledge in theory, at others it is for practical experts with great skill in application. When the first is required, the school is usually of collegiate grade, and is referred to as an institution of higher technical education. Where the demand is for trained workers, the school may be of the secondary level, and may be a trade or industrial school.

The Continent Assumes the Lead in Higher Technical Education.—France and Germany took the lead in establishing higher technical schools. In America, the demands of the new and changing society which have been set forth in Part I of this volume gave rise to higher technical institutions. The first of these appeared about 1820, and by 1870 the movement was in full swing.

Italy, perhaps, first held supremacy in engineering education in modern times, but France soon became the recognized leader in this field, particularly in the training of military and civil engineers.

Before 1790 the Ecole des Mines was established in Paris, and in 1795 the Ecole Polytechnique de France was opened. The government established this engineering school as a source from which civil and military engineers could be recruited. Bache, in his "Report on Education in Europe to the Trustees of Girard College for Orphans" (1836), as reported in Bennett's "History of Manual and Industrial Education," page 349, gives a summary of the curriculum which includes pure and applied mathematics, physics, chemistry, architecture, drawing, French, and German.

As a response to the demand for leaders in the developing manufactories, the Ecole Centrale des Arts et Manufactures was founded in Paris in 1829. This school emphasized applied science rather than pure science. Courses were given in the "steam engine" and in "railroads"—Bache notes much in the way of laboratory work of a practical nature in physics and mechanics. Excursions for the embryonic architects to incompleted buildings and finished buildings were featured. The school received many foreigners.

Germany, also, established such schools. Those of Vienna, Karlsruhr, Munich, and Freiberg were important. These, too, featured the laboratory method.

Higher Technical Education in the United States.—The Americans were not long in establishing advanced technical schools. The basis of the whole mechanics' institute and lyceum movement was the teaching of science, and this impulse showed itself also in the field of higher technical education. Hence there appeared institutions which took students for full-time work, and trained them for superior positions in agriculture, the mechanic arts, and engineering.

Perhaps the first American technical school of collegiate grade was the Gardiner Lyceum in Maine, founded in 1823. This was, in part, a manual labor school, and boasted of its large and commodious shops of lathes, but for the most part it was a full-time scientific school. However, it did offer short-time course. Benjamin Hale was its principal; he later became prominent as president of Hobart College.

Bennett in his "History of Manual and Industrial Education," page 348, quotes the purpose of this institution as given in the inaugural address of the principal—"to give instruction in those branches which are most intimately connected with the arts, and to teach them as the foundation of the arts. . . . They (the students) must be taught the application of the laws (of the sciences). They must be made acquainted with machines."

The curriculum stresses the application of the principles of physics, chemistry, and mathematics. Agriculture, navigation, civil engineering, and mineralogy appear in the curriculum. A statement concerning the methods of instruction is given in Bennett (page 349) wherein the school sets forth its method as one of practical application as well as recitation room discussion. The laboratory method on actual jobs, it is noted, is employed so that the pupil may become familiar with those processes which will be necessary to him in actual life.

The funds by which the Gardiner Lyceum was supported came from gifts, tuition fees, and contributions from the state legislature. After some ten years the legislature withdrew its support, and the school closed.

No doubt the most important of these higher technical schools was the Rensselaer School at Troy. This is today the Rensselaer Polytechnic Institute, admittedly one of the best engineering schools of the country.

The changing and developing industrial society called into being this school, whose curriculum was anything but traditional. Palmer Ricketts, in the "History of Rensselaer Polytechnic Institute," page 9, says that "it aimed to benefit the sons and daughters of farmers and mechanics in the application of science to the common purposes of life." Also, "in the application of experimental chemistry, philosophy, and natural history, to agriculture, domestic economy, the arts, and manufactures."

Stephen Van Rensselaer, a wealthy landowner of New York State, and a Harvard graduate, founded the institution in 1824. He held several important positions in the state, among which were the presidency of the board of agriculture and membership in the Erie Canal Commission. In this work he became acquainted with one Amos Eaton, a pupil of Professor Silliman, a Yale professor of chemistry, geology, and mineralogy. Van Rensselaer employed Eaton, and sent him out upon a lecture tour through the Erie Canal region for the purpose of raising the standard of education among the farmers. The lectures were on the application of scientific principles, and the plan worked so well that Van Rensselaer founded his school, with Eaton as senior professor.

A number of neighboring farms and workshops entered into agreements with the school, and provided places for the students to apply the principles taught them in the school. Van Rensselaer evidently did not believe in the Fellenberg plan of combining manual labor with school studies. A special provision for the training of teachers of the applied sciences was formulated by giving practice in lecturing to a number of selected students. Supervised physical exercise was another feature of this institution.

Roy Palmer Baker in "A Chapter in American Education" states that, because of the progressive character of the work and the high standing of its professors, Rensselaer drew many college graduates, and at times almost one-half of the enrollment were college graduates. Research was carried on in a way remarkable for that time. It might be said, according to Baker, that Rensselaer was the first graduate school in America.

Following the example of these two schools, the scientific course was opened at Union College in 1845, the Sheffield School at Yale in 1827, and the Lawrence Scientific School at Harvard in 1847.

After the first ten years of its existence, the character of the Rensselaer school gradually changed until it became a college of engineering, in the accepted sense of the word. The need of the new country for engineers was at the root of this change. In 1835 the department of mathematical arts graduated four men with a new degree—"Civil Engineer" (as opposed to "Military Engineer"). These men were qualified in "engineering and technology."

The Teaching of Scientific Agriculture in America.—The agriculture societies and the college professors who offered science courses applied to agriculture were not the only agencies seeking to develop agricultural education. Legislation was a third center of effort acting for progress.

Illinois was one of the states which led in interest in agricultural education. In 1850 the president of the Illinois State Teachers' Institute, Jonathan Turner, made an address in which he pleaded for a state industrial university which should educate for all agricultural and industrial occupations. This was the first step, it appears, in the movement that established the great state land-grant universities of this country. Turner again presented his views at a farmers' convention in Putnam County, Illinois. A committee adopted resolutions embodying his principles, and planned to advertise these ideas. A state convention at Springfield was held for this purpose, where the classical and theological professors jeered at the plan to establish such an institution as Turner had in view. Turner presented a memorial to the convention in which he suggested that an appeal be made to Congress for support. Many educators, farmers' organizations, journals, and societies assisted in aiding the cause.

In March, 1854, the resolutions of Turner which had passed the Illinois legislature were presented to Congress. In April,

Mr. Yates, a congressman and a personal friend of Turner, asked Turner to prepare a bill for Congress. Unfortunately Yates was not returned to Congress, and a growing opposition to land-grants on the part of the eastern states necessitated a delay. Senator Trumbull of Illinois assured Turner that a delay might be beneficial to the cause.

Morrill and the Land-Grant Colleges.—Justice Morrill became a Representative from Vermont in 1855. Trumbull sent him all the literature, documents, and papers pertaining to the land-grant idea for universities, and asked him to introduce a bill for the necessary legislation. This he did in December of 1857. Morrill proved to be a vigorous champion of the measure, and the bill passed the House, but failed in the Senate. It was reintroduced in 1859 and passed both House and Senate, but President Buchanan vetoed it.

In the campaign of 1860 both Lincoln and Douglas pledged support to the measure. When Congress met in 1862, Morrill again presented the bill. It passed both Houses, and became a law when Lincoln signed it in July, 1862.

This was one of the most important pieces of educational legislation the country yet had. Edmund James in his tract "The Origin of the Land-Grant Act of 1862" (University of Illinois, 1910) says that it was the most comprehensive "scheme for the endowment of higher education ever adopted by any civilized nation."

By the provisions of the Land-Grant Act, 30,000 acres of public land per senator and representative in Congress were granted to provide colleges of agriculture and the mechanic arts in the several states. This was the root of our present state colleges of agriculture and the mechanic arts, and many of the state universities.

Engineering Education After the Civil War.—Mechanical engineering was not early in joining its brethren, military engineering as given at West Point and civil engineering as given at Rensselaer. Not until after the Civil War was provision made for mechanical engineering. By 1866, however, Rensselaer, Massachusetts Institute of Technology, and the Polytechnic College of Pennsylvania offered courses leading to the M. E. degree.

The great inventions of Fulton, Whitney, Howe, and Corliss created a demand for mechanical engineers which is perhaps yet unsatisfied.

The Worcester County Free Institute of Industrial Science (later the Worcester Polytechnic Institute) was founded at Worcester, Massachusetts, in 1865. John Boynton pledged $100,000 for the endowment of a free school, if the citizens of Worcester would provide land and buildings. This they did, and the school was opened. Later in the year, Ichabod Washburn provided $25,000 and later $50,000 for the establishment of a machine shop as a part of the institution. Washburn's idea, as expressed in the *Journal* published by the Worcester Polytechnic Institute July, 1906, included the teaching of the trade with instruction in the theory underlying the science of that trade.

Washburn placed the work of developing his idea in the hands of Milton Higgins (1842–1912) who firmly believed that a workman, whether a prospective engineer or machinist, should be trained in a productive shop. The "real" manufacturing shops of the Worcester school stand as evidence to this.

Thus a new type of engineering school was established. It consisted of a mechanical engineering course which included theory and practice. The former was provided in the courses in applied science, while the latter was obtained in a shop run upon a commercial basis, producing articles to be sold on the market. The shop work was organized with a definite end in view; it was a very real substitute for apprenticeship.

Part Three

THE CONTRIBUTIONS OF SOCIAL AGENCIES AND EDUCATORS TO VOCATIONAL EDUCATION

Part Three deals with the contributions to vocational education made by the various social and philanthropic agencies. The value and importance of some of these attempts have recently been over-emphasized in some circles, but it is reasonable to believe that in most cases the spirit in which these attempts have been carried on has been flavored with deep sincerity and the desire to provide a much needed relief.

In this section a brief mention must suffice in many cases rather than a detailed study of the various institutions.

The work of philanthropical societies, social associations, fraternal orders, organized labor, scientific societies and the like has been considered. Finally the contributions of modern movements in education and of the Federal Government have been listed.

SOCIAL AGENCIES AND VOCATIONAL EDUCATION

Introduction.—Societies for the promotion of vocational education have set forth the results of painstaking research, and have sought to bring before the public the needs, methods and plans for vocational education. Organized labor has constantly referred to the subject of trade and industrial training. Chambers of commerce as well have adopted resolutions voicing the demand for education of an industrial nature. And, not the least, business itself has set up plans for the training of its workers. The purpose of these pages is to note briefly some of the motives expressed by these social agencies which have fostered vocational education.

The Massachusetts Commission on Industrial and Technical Education.—With the opening of the twentieth century there was manifested a tendency to emphasize the importance of vocational education as an end in certain school activities. A most influential agency in inaugurating a definite movement for vocational education in both general and special schools was the Massachusetts Commission on Industrial and Technical Education, appointed in 1905 to investigate the needs for education in the different grades of skill and responsibility in the various industries of the state. Its report, issued the following year, is replete with material of aid to vocational training. It gives a review of conditions giving rise to the vocational movement.

Prominent in the report is a pertinent paragraph on manual training, which is quoted in Anderson's "History of Manual and Industrial School Education," page 199:

"The wide indifference to manual training as a school subject may be due to the narrow view which has prevailed among its chief advocates. It has been urged as a cultural subject mainly useful as a stimulus to other forms of intellectual effort,—a sort of mustard relish, an appetizer, to be conducted without reference to any industrial end. It has been severed from real life as completely as have the other school subjects."

The change in methods of manufacture is an additional factor in the trend toward vocational education, according to the Commission's report. The decay of apprenticeship, which increases local and international competition, is also mentioned as a contributing item. Besides these general considerations the Commission noted a further reason in the fact that "the lack of skilled workmen is in many industries making the process of manufacture difficult and expensive." The report continues,

"This lack is not chiefly a want of manual dexterity, though such a want is common, but a want of what might be called industrial intelligence—mental power to see beyond the task which occupies the hands for the moment to the operations which have preceded, and to those which will follow it.

"Manufacturers believe that a system of industrial education wisely planned would tend to develop such intelligence, while it increased technical skill." (Ibid, p. 200.)

From Anderson's "History of Manual and Industrial School Education," page 201, is the following critique:

"A further consideration, and one to which the Commission was, perhaps, the first to direct general attention, was the large number of boys and girls, who, after completing the period of compulsory school attendance at fourteen, and before being admitted to industrial apprenticeship at the minimum age of sixteen, become permanently fixed in some blind alley occupation. A careful investigation conducted by the Commission's Subcommittee on the Relation of Children to the Industries reported that sixty-eight per cent of the children who commence work between fourteen and sixteen are subjected to the evil influences of these unskilled industries or are in mills. They have wasted the years as far as industrial development is concerned, and in many cases they have forfeited the chance ever to secure it, because of lack of education."

The Commission recommended provision through the existing public school system and independent public schools for industrial education directly available for vocational purposes. For the public schools it recommended instruction and practice in the elements of productive industry, including agriculture and the mechanic and domestic arts. It recommended further that towns and cities be permitted to "provide independent schools for instruction in the principles of agriculture and the domestic and mechanic arts, also evening courses for persons already employed in trades, and instruction in part-time classes of children between the ages of fourteen and eighteen who may be employed during the remainder of the day." Finally the report recommended the appointment of a commission on industrial education. The recommendations of the commission were formulated into a bill which was passed by the Massachusetts Legislature in 1906. Thus, the "Douglas Commission" report attracted widespread attention, and its influence has been greatly felt throughout the entire United States.

The Report of the Committee on the Place of the Industries in Public Education of the N. E. A.—The National Education Association, acting through its committee on the Place of the Industries in Public Education, appointed in 1907, shows the trend of educational thought toward vocational training. General dissatisfaction with manual training is expressed definitely in the report of the National Education Association committee. The report asserts:

"In intermediate schools, industrial occupations are an important element in the wide range of experience necessary for the proper testing of children's aptitudes as a basis for subsequent choice of specific pursuits either in vocations or in higher schools" and "In secondary schools, industrial occupations properly furnish the central and dominant factor in the education of these pupils who make final choice of an industrial vocation" (Report of the Committee on the Place of the Industries in Public Education, 1910, pages 4 and 6.)

Handwork, the committee insists, must be more than a "general culture" subject; it must furnish a large proportion of the pupils a definite preparation for earning a living. The committee believed that in small towns the manual training courses can be modified so as to include practical industrial education; but for larger centers it advocated the establishment of vocational industrial high schools. These reports suggest the motive power furnished by the wealthy business interests, which support industrial training as an efficient plan in recruiting skilled workers.

The Meeting of the National Association of Manufacturers in 1907.—At the gathering of the National Association of Manufacturers of the United States in 1907 a report on industrial education was submitted. The report, as quoted on page 205 of Anderson's "History of Manual and Industrial Schools," stated: "Officers and managers of the principal industrial and trade schools through the country ... agree with us that a much higher-grade mechanic can be graduated from a trade school than can be produced through the apprenticeship system in the old way."

The National Society for the Promotion of Industrial Education.—The National Society for the Promotion of Industrial Education was organized in 1906 to unite the many forces making toward industrial education in the country.[1] Industrial education to this society means "that schooling which deals with training of direct vocational value to the industrial worker." The constitution of the society contains the statement of aim: "To bring to public attention the importance of industrial education as a factor in the industrial development of the United States." Its roster contains members from manufacturers, educators, and social workers. Its annual meetings and its surveys have been a positive influence in the origin and development of an industrial education.

Labor and Vocational Education

Organized Labor a Group Which Has Influenced Education.—Organized Labor has been a group in American life

[1] Dr. F. Theodore Struck's book, "Foundations of Industrial Education," (John Wiley Sons, 1930) contains the most authentic and detailed history of the National Society for the Promotion of Industrial Education available. This society is now known as the American Vocational Association.

that has had much to do with the development of trade and industrial education.

The 1920 Convention of the American Federation of Labor.—The 1920 Convention of the American Federation of Labor instructed the Executive Council, "in order to make easily available Labor's position on education, and to add to the effectiveness of Labor's contribution to educational progress, to have compiled and published in pamphlet form the official declarations of the labor movement upon education." The result was the pamphlet, "Education for All" published by the Federation in 1922. Upon the first page of this document there appears the following statement:

"This foreword should call attention to the fact that from the beginning of the labor movement, and to the extent to which they could make known their aspirations it has been ever the workers who have advocated and advanced the cause of education for all." Also the following: "Wherever and whenever they (the workers) become articulate through organization, the workers have demanded and forwarded their hope for a better opportunity for the coming generations through access to educational facilities for all the children of our country, as well as for adults who have been denied them in their youth."

Labor Considered the Problem of Trade Training at an Early Date.—The convention of 1881 was the first convention of the American Federation of Labor. This meeting declared:

"We are in favor of the passage of such legislative enactments as will enforce, by compulsion, the education of children; that if the state has the right to exact certain compliance with its demands then it also is the right of the state to educate its people to the proper understanding of such demands." ("Education for All," page 3.)

The conventions of 1888, 1894, and 1911 reiterated these sentiments, and resolutions of a similar nature were adopted. Since 1918 almost every convention has incorporated in its educational platform the following resolution, "Better enforcement of compulsory education laws, and the establishment of a minimum school leaving age of 16 years." With the plan of a sixteen years minimum age for a youth to enter industry, there grew up plans for changes in traditional school activities to accommodate these older pupils.

Labor and Adult Education.—The American Federation of Labor has encouraged adult education. In 1918, the convention expressed the following sentiment: "The wider use of the school plant securing increased returns to the community through additional civic, social, and educational services to both adults and children is desirable." ("Education for All," page 9.)

Labor and Federal Authority for Education.—Labor's interest in education is further noted by a resolution passed at the 1918 convention. This resolution recommended "the establishment of a Federal department of education, headed by a cabinet officer." ("Education for All," page 9.)

Labor and the Evening School.—The 1911 convention declared in favor of public night schools for children over 16. In 1915, and following, the recommendation was broadened to include "night schools for all persons over 18 years of age desirous of further educational opportunities, either cultural or vocational." ("Education for All," page 11.)

Labor and Part-Time Education.—Continuation schools were mentioned at the 1911 convention. The pamphlet "Education for All," page 11, states:

"In 1911 organized labor asked for legislation emphasizing the necessity for continuation schools, both of the part-time day type for the younger boys and girls and of the evening type for more mature workers, and for the all-day type trade preparatory school for boys and girls between 14 and 16 years of age. In 1915 the recommendation was endorsed for compulsory day time continuation schools for all children not high school graduates in industry between the ages of 14 and 18 years, for not less than five hours per week at the expense of their employers. In 1918 the convention adopted the following statement, including a model part-time education law for state use, which has been reendorsed by following conventions: 'Our country stands badly in need of judicious practical part-time education and part-time employment state legislation, and the organized labor movement must take the initial step in this direction. Some of our states, such as Pennsylvania, New York, Indiana, and Wisconsin, have already on their statute books some form of legislation intended or expected to help or control children who have reached certain ages or certain school grades and who contemplate undertaking some employment. It is our belief that a model state part-time law should be prepared and urged for enactment by the several state legislatures at the earliest possible date.'"

Labor's Repeated Utterances for Efficient Industrial Training.—The attitude of the Federation toward industrial education is evidenced in a declaration of the 1907 Convention:

"We favor the best opportunities for the most complete industrial and technical education obtainable for prospective applicants for admission into the skilled crafts of this country, particularly as regards the full possibilities of such crafts, to the end that such applicants be fitted, not only for all usual requirements, but also for the highest supervisory duties, responsibilities, and rewards; and the Executive Council is directed to give this subject its early and deep consideration, examining established and proposed industrial school systems, so that it may be in a position to inform the American Federation of Labor what in the Council's opinion would be the wisest course for organized labor to pursue in connection therewith." ("Education for All," page 12.)

In 1908, the following is culled from the convention minutes:

"Industrial education is necessary and inevitable for the progress of an industrial people. . . . Organized labor has the largest personal interest in the subject of industrial education, and should enlist its ablest and best men in behalf of the best system, under conditions that will promote the interests of the workers and the general welfare."

"Education for All" (page 13) quotes the 1909 convention report as saying:

"The subject of education, industrially, concerns not only the wage earners but every inhabitant of the nation. It is, therefore, necessary and eminently proper that it be administered by the same authority and agency which administers our public school systems and such other institutions as are concerned in public welfare. Already reference has been made to the false position in which some elements of employers would place our movement upon this subject. All we ask of fair-minded men is a comparison of the utterances of our opponents with our own. We contend that education in America must be free, democratic, conducted by, of, and for the people, and that it must never be consigned to, or permitted to remain in the power of, private interests, where there is sure to be danger of exploitation for private profit and wilful rapacity. Under the pretence of industrial education private agencies for personal profit have perverted the term, resulting in a narrow and specialized training, to the detriment of the pupils, the workers and the people generally. . . .

"The demand for supplemental technical instruction is measured by the necessity for training in particular trades and industries. The chief aim of such instruction should be to present those principles of arts and sciences which bear upon the trades and industries, either directly or indirectly. The economic need and value of technical training is not to be disregarded, and cognizance should be taken of the fact that throughout the civilized world evening and part-time day technical schools enroll twenty pupils to every one who attends the other types of vocational schools. There should be established, at public expense, technical schools for the purpose of giving supplemental education to those who have entered the trades as apprentices.

"We favor the establishment of schools in connection with the public school system, at which pupils between the ages of 14 and 16 may be taught the principles of the trades, not necessarily in separate buildings, but in separate schools adapted to this particular education, and by competent and trained teachers. The course of instruction in such a school should be English, mathematics, physics, chemistry, elementary mechanics and drawing, the shop instruction for particular trades and for each trade represented, the drawing, mathematics, mechanical, physical and biological science applicable to the trade, the history of that trade, and a sound system of economics, including and emphasizing the philosophy of collective bargaining. This will serve to prepare the pupil for more advanced subjects, and, in addition, to disclose his capacity for a specific vocation. In order to keep such schools in close touch with the trades, there should be local advisory boards, including representatives of the industries, employers and organized labor. Any technical education of the workers in trade and industry being a public necessity, it should not be a private but a public function, conducted by the public and the expense involved at public cost."[2]

EFFICIENT INDUSTRIAL TRAINING

To continue this series of quotations, there is one from the 1910 convention which states clearly the position of the Federation:

"Conservation is one of the topics uppermost in the mind of the American people today, but there is one phase of conservation which is not receiving the attention it deserves; I refer to the conservation of brain and brawn of American youth. Our school systems are giving only a one-sided education; the boy may go to school and prepare himself for professional or commercial life, or he may drop out of school and enter a trade with no particular preparation and become a mediocre workman. Training of brain and muscle must go together for the complete preparation of men. While the public schools and colleges aim only at teaching professions, the greatest need of America, educationally, is the improvement of industrial intelligence and working efficiency in the American youth. We need an educational uplift for the work of the boy who will work with his hands, and we not only need to give an educational uplift to craftsmanship, but the school needs the help of the workman and his better work in education." (1910 Convention report, pp. 40 and 273.)

The 1911 convention favored the appointment of a national commission preliminary to securing vocational education legislation.

The convention of 1914 declared valuable and comprehensive the report of the national commission appointed by the President on federal aid to trade and vocational education.

In 1915 the convention declared,

"In connection with the subject of industrial education and vocational training, we submit that the Federal Government should afford generous financial aid in this matter fraught with so much value to the workers, to the people generally, and to the stability of our country."

The same convention voiced the following sentiment:

"Then, too, industrial education should not be allowed to coordinate itself with any arrangement which will bring trained and experienced workers into any trade without regard to the demand for labor in that particular trade or calling. A proper apportionment of the supply of labor to the demand for labor must be maintained. What good will industrial education serve, what benefit can be derived, if by such teaching we are to produce a greater number of trained and skilled workers than is required or can possibly be employed in the respective trades or callings? Industrial education under such conditions can only increase the existing economic pressure on the workers. Industrial education must, therefore, be based on a careful survey of industrial conditions and trade requirements, and should meet the needs and requirements of the workers as those of employers and of the industry." (Quoted in "Education for All," page 14.)

The following is quoted as part of Labor's reconstruction program after the close of the World War:

"It is also important that the industrial education which is being fostered and developed should have for its purpose, not so much training for efficiency in industry, as training for life in an industrial society. A full understanding must be had of those principles and activities that are the foundation of all productive efforts. Children should not only become familiar with tools and materials, but they should also receive a thorough knowledge of the principles of human control, of force and matter underlying our industrial relations and sciences. The danger that certain commercial and industrial interests may dominate the character of education must be averted by insisting that the workers have equal representation on all boards of education or committees having control over vocational studies and training." ("Education for All," p. 15.)

Labor and the Smith-Hughes Act.—The American Federation of Labor in the 1918 convention highly commended the Smith-Hughes act and expressed pride in the fact that, as a group, labor played a part in the cooperative drive to push through such legislation.

Labor's Investigation of Educational Activities Under Union Auspices.—In accordance with instruction from the 1918 convention, a special committee conducted an investigation of the educational activities under union auspices; and of the relationship of such activities to the public schools. The 1919 convention adopted the following general conclusions, quoted in "Education for All," page 19:

"In the judgment of your committee the most important differences in the system described are in the varying degrees of cooperation of the unions with the public schools and of the unions with each other. In New York City it is chiefly one large international, The Ladies' Garment Workers, which has developed its own educational department and secured cooperation with the public schools to the extent of the use of four elementary school buildings for their unity centers, one high school for their central classes, and the services of teachers of English. All of the educational work of The Ladies' Garment Workers in Philadelphia is in cooperation with the public schools. In Boston, the Central Labor Union has a high school building. In Chicago, also, the Chicago Federation of Labor, in conjunction with the Women's Trade Union League, has organized the educational work, the public schools furnishing a large proportion of the teachers and meeting places. In Los Angeles, while the movement was initiated by the unions, the Board of Education now has full control, though utilizing the close cooperation of the unions in working out courses and methods.

"It is unnecessary to emphasize labor's deep appreciation of the value of education. And it is unnecessary to more than mention organized labor's pride in the part it took in the establishment of our public schools and the consistent and vigorous stand it has taken ever since for the highest development of our system of public education."

"One of the things that impressed the committee in the classes of The Ladies' Garment Workers in New York City was the feeling of the students that the classes belonged to them, that they were at home in them, and took a collective pride in them. That is high praise for those classes, but it is also an indication of a serious shortcoming in our public schools, and in the attitude of the public, that is not limited to New York City. For that sense of part-ownership should be in the minds of the students in all public classes.

"Where the types of courses and instruction desired cannot be obtained from the public schools, we believe that all interested unions, working through their central labor bodies, should cooperate in organizing their educational work."

In 1920 the convention directed a special committee to cooperate "in studying the possibility of coordinating the present educational institutions and activities conducted under the auspices of organized labor."

The Future and Labor's Part in the Vocational Movement.—In the future it appears that organized labor, so fundamental a force in American industry, can be a significant factor in the determination of the policies and practices of trade and industrial education. It is obvious that progressive labor leaders are giving more serious thought to the question of training for vocational efficiency.

The Y. M. C. A. and Vocational Education.—The Young Men's Christian Association has done much to stimulate thought and activity in the field of vocational education. The yearly reports of the Association give ample indications that it makes a serious attempt to assist young workers in their trade and industrial education. The conferences and round tables of the association invariably include the topic of industrial education.

The Early History of the Y. M. C. A.—The history of the development of the Y. M. C. A., and in particular its relationship to vocational education, is an example of progressive service and a picture of an attempt at the partial solution of trade training. What has been true of the Y. M. C. A. is also true of the Knights of Columbus, the Young Men's Hebrew Association and the Young Women's Christian Association. These younger agencies are following the example of the older association, and their histories are, for the most part, similar to that of the Y. M. C. A. The only difference appears to be that they are serving different social groups and are working in a more limited area.

In 1844 the first Y. M. C. A. was organized in London. Seven years later, in 1851, two branches were organized in North America. One was started in Montreal, the other in Boston. Educational work was discussed at a meeting in London in 1845. A committee reported the organization and conduct of a series of lectures during the fall and winter of that season. These lectures had been held in Exeter Hall (page 20, "Association Educational Work," George B. Hodge, Association Press, 1912). This was the first educational venture of the Association. The Exeter Hall lectures continued for twenty years.

Y.M.C.A. AS AGENCY OF VOCATIONAL TRAINING

The Y. M. C. A. Inaugurates Classroom Activities.—The desire to assist men in their religious development and daily life led the Y. M. C. A. to inaugurate classroom activities. From 1851 to 1866 the Y. M. C. A. established itself in North America. Reading rooms were, perhaps, the first concrete evidences of the Association's interest in education.

In the fall of 1892 Mr. W. H. Coughlin, who for a year or two had been teaching free-hand drawing and one or two other subjects in the Association in Brooklyn, New York, was asked by Mr. Edwin Lee, the general secretary, to give half of his entire time to the promotion of educational class work, lectures, and library activities. For two years (1892–1894) he gave half time to this duty, and afterwards devoted full time to work in education. In 1906 he joined the ranks of the emeriti. Thus began the educational secretaryship in the Y. M. C. A.

The Y. M. C. A. Actively Engages in Trade Training, 1892.—In 1892 at Springfield, Ohio, W. J. Fraser, the general secretary, and D. F. Graham, a skilled mechanic, both with a conviction that the Y. M. C. A. should help men in their industrial pursuits, conducted courses in pattern making, tool making, and cabinetmaking, all supplemented by the appropriate mechanical drawing and shop mathematics. The work was carried on in a foundry and blacksmith shop two blocks away from the Y. M. C. A. building. Such was the beginning of shop work and vocational training in the Y. M. C. A.

Stimulated by these endeavors other branches followed suit. The branches of Hartford, Connecticut; Chicago, Illinois, and New York City, New York, were among the early pioneers.

A Federal Report Mentions the Y. M. C. A. as an Agency of Vocational Training.—The Twenty-fifth Annual Report of the Commissioner of Labor (1910) states the following (pages 22–23):

"The Young Men's Christian Association schools are the most widely distributed class of philanthropic agencies for industrial training, there being about 180 associations scattered through the country which give industrial, scientific, technical, and trade instruction to a greater or less extent."

"The first automobile school in America was instituted by the Young Men's Christian Association in Boston in 1900. There are now 37 such schools with over 3,000 pupils, many of whom are owners of machines. The first apple-packing school in America, so far as known, was organized by the Y. M. C. A. in North Yakima, Washington, in 1907. A three weeks' agricultural school for 240 farmers was held by the Mount Pleasant, Iowa, Y. M. C. A. in 1907, and has been repeated annually since. . . . The poultry school of Portland, Oregon, with nearly 100 boys and young men, has been conducted for two years in cooperation with the state agricultural college." (Ibid—page 365.)

The Present Activities of the Y M. C. A. in Industrial Education.—Since that time the Y. M. C. A. has entered almost every field of vocational education. Its work for soldiers and sailors, colored men and boys, and railroad workers as well as for all sorts and conditions of young people is well known. It has established night schools, employment offices, and vocational guidance interviews in many of its branches. The Y. M. C. A. has, of course, often stressed religious and "character" development rather than purely industrial training. Informal educative agencies have been developed to a remarkable degree. Discussion groups, forums, vocational clubs, and lectures have been of much assistance in preparing the way for the richer development of many young men vocationally. At present, the Y. M. C. A. is pioneering in the fields of aviation and the semi-professional occupations. Also much thought is being given to the study of the junior college.

A few significant figures will indicate the scope of the Y. M. C. A. work. These figures have been taken from the "Report for the Meeting of the Board of Governors, Chicago, 1929." They are based on the "Year Book Data," 1928–1929.

	Number of Students Enrolled
Accounting	2,937
Blueprint Reading	672
Algebra	2,549
Architectural Drafting	691
Business English	2,086

PRIVATE AGENCIES

	Number of Students Enrolled
Electricity	655
Foremanship	1,100
Shop Mathematics	867
Stenography	1,444
Traffic Management	765
Radio	1,777

(These represent only a few of the many courses.) The "1929 Yearbook," page 144, lists the following figures for the United States and Canada.

Number of Students Enrolled	71,800
Number of Instruction Hours	474,989
Number of Vocational Interviews	47,482

Private Agencies Which Encourage Vocational Education.—In addition to the Y. M. C. A. and similar organizations, private agencies have provided means for vocational training.

Philanthropic industrial schools are most diverse in character. Some were established for general training, and have since added industrial departments. Others were founded as trade schools. Some are maintained from original endowments. Others are aided by the state, city or individuals. Some charge a nominal tuition fee, others are absolutely free. Some are trade preparatory schools, some trade extension schools. Nearly every form of industrial training for both sexes is provided by schools of this class.

The nature of this study precludes any detailed discussion of the various schools of this type. Several representative schools of this sort include the Williamson Free School of Mechanical Trades, Delaware County, Pennsylvania; Girard College, Philadelphia, Pennsylvania; Carnegie Technical Schools, Pittsburgh, Pennsylvania; Pratt Institute, Brooklyn, New York; David Ranken, Jr., School of Mechanical Trades, St. Louis, Missouri; the Arsenal Technical High School, Indianapolis, Indiana; Mechanics Institute of Rochester, New York, Hebrew Technical Institute, New York City; New York Trade

School, New York City; and the Illinois Manual Training Farm, Glenwood, Illinois.

Schools of this sort have done great work in keeping alive the practices of industrial training. They are usually dependent upon private endowments, and, hence, their contributions to education have been local and limited in scope.

As an example, the National Trade School of Indianapolis, Indiana, will be described. Agitation for the establishment of a trade school at Indianapolis was begun in 1903. In 1904 the grounds of the United States arsenal were purchased with funds raised by popular subscription from the citizens of Indianapolis. In April, 1904, the Winona Technical Institute was incorporated. The name has since been changed to the Arsenal Technical High School.

The school had for its purpose the teaching of the trades. Realization of the importance of real shop work and training prompted the school authorities to confer regularly with advisory boards of the various national employers associations. Special arrangements with many industries have been reached whereby the students are given the opportunity of spending part of their time working in commercial shops. In 1909 the school went into the hands of a receiver and underwent a complete reorganization and reconstruction so that it might enjoy public aid. The trades taught include lithography, printing, molding, machinist, tile setting, carpentry, painting, paper hanging, pattern making, chemistry, and pharmacy. The unions have supported the reorganized school on the condition that neither employers nor workmen involve the school in disputes between capital and labor, and that employers do not use the school as a strike-breaking institution.

The Inadequacy of Private Endowment.—It is interesting to note that privately endowed schools in many cases find that original endowments, which at the time of establishment seemed ample to provide first-class equipment and still care for the maintenance and annual running expenses, fail to appear ample after a decade or two. Two or three hundred thousand dollars in the eighteen-nineties could be used to great advantage; the same amount today would be entirely inadequate for the same purposes.

Fraternal Association and Vocational Education.—The various fraternal societies, such as the Masons, Elks, and Loyal Order of Moose, have indulged in educational activities from time to time. These activities have, in most cases, been limited to the children of members of the respective associations.

The Masonic Order and Vocational Education.—The Masons became interested in education at an early date. The Grand Lodge of New York claims a share in the honor of having helped to found the public school system in New York. In 1809 it adopted a scheme whereby each of the twenty-two lodges in New York City were to contribute ten dollars per year toward the education of fifty poor children whose fathers were, or had been, Masons. A school committee was appointed. This committee worked with the trustees of the New York Free School of whom Grand Master De Witt Clinton was chairman. This work continued until 1818, at which time the free schools were firmly established under the supervision of the state.

The Masons and the Public Schools.—Throughout its history in America, Freemasonry has stressed support of the public school system. Various studies have been undertaken occasionally by the Masons to bring before the members the importance of public education. Bulletin No. 8 of the Masonic Service Association of the United States is devoted to "The Needs of Our Public Schools."

George Washington University Endowed by the Masons. —The donation of one million dollars to George Washington University by the Supreme Council of Scottish Rite Masons, Southern Jurisdiction, is an indication of the interest of Masons in education. This gift was made in 1928 for the establishment of a school of government at that university as a memorial to George Washington, the Mason. From statements made by the leading Scottish Rite officials, it is specifically understood that the educational program and aim of the Supreme Council is a primary education in a public school for every American boy and girl, and a great National University at Washington supported by the Federal Government.

Masonic Loan Funds for Students.—Loan funds for college students have been another feature of Masonic interest in education. There is such a fund at Oregon Agricultural College. Alabama Lodges have a Knights Templar Educational Loan Fund which in 1928 provided about $30,000 to some sixty students.

The Masonic Educational Endeavors at Utica.—From the point of view of vocational education the Masonic Home and Hospital of the Grand Lodge of New York at Utica is of interest. After much study and discussion, the Masons of New York decided to provide for needy Masons and their families through the establishment of the famous Utica plant. The first unit of this plant was dedicated in 1892. Other units followed in 1907, 1910, 1917 and 1922. The primary interest of the home is to provide food, shelter, and clothing for needy children of widowed Masons. After a time the school division was discontinued and the children were sent to the public school. The brochure of the "Masonic Home and Hospital," page 4, states, "Under the new method an effort is made to give each child whatever in the way of education that child is able to assimilate and put to good use. The result is that many of our children have gone through college and entered professions, while others fill perhaps humbler positions but fill them no less creditably." At present plans are being drawn for the reorganization of the school with provision for shop training.

The Thomas Rankin Patton Institution, a Sample of Masonic Interest in Education.—The Pennsylvania Masons have a vocational school for boys who are sons of Masons. This is the Thomas Rankin Patton Masonic Institution for Boys. The plant consists of an administration building, a school building, a dormitory, and the shops. Facilities are provided for some sixty boys. The object of the institution is to provide experience in brick laying, carpentry, and machine-shop practice. The 1928 report tells of the graduation of nine boys.

ORDER OF MOOSE AND EDUCATION

The Elks and Vocational Education.—The Elks do not, as a body, carry on any extended work along educational lines. Children of deceased members are often aided by individual lodges, which often use the public schools as the agency of education.

The Order of Moose and Education. Mooseheart.—One of the most elaborate of the privately endowed institutions giving vocational training is that at Mooseheart, Illinois, conducted by the Loyal Order of Moose. Mooseheart is a vast, park-like estate of over one thousand acres, situated about thirty-five miles west of Chicago. More than 175 cottages, dormitories, workshops, and administrative buildings comprise the physical equipment. About two thousand orphan children are cared for in this city devoted to child raising. They are given a home, a high school education, and instruction in a trade. They range in age from infancy to eighteen years. The estimated value of the plant is about ten millions of dollars, and the income each year from the members of Moose throughout the country is over one million dollars.

Mooseheart owes its existence to James J. Davis, Secretary of Labor in the cabinets of Presidents Harding, Coolidge and Hoover. He, as Supreme Director of the Moose, has been given full credit for this great educational endeavor. The school is supported by annual sums taken from the dues of the brothers in Moose.

The academic course of study at Mooseheart is approved by the leading higher institutions of the vicinity, which permit the graduates to enter their organizations by certification of graduation. In vocational training forty courses are offered, designed to give a foundation upon which the student may continue to build. Verne A. Bird is at present director of education, while Ernest Roselle is superintendent. The importance of this experiment is becoming fully recognized, and, it is said, over 150,000 visitors, many of whom are educators, enter the gates of the city yearly.

A quotation from an article by Superintendent Roselle in the "Mooseheart Magazine" of August, 1929, page 27, is perhaps an indication of the plight of so many privately endowed institutions:

"In the field of education we have an opportunity equal to that in home training. Mooseheart is already widely regarded as the champion of a practical education for childhood—an education that includes parallel and related training of the hand, the heart, and the mind. We have gone far; but much must still be done before we can realize our full objectives. Again, the factor holding us back is the physical plant. We have a well defined philosophy of education. We have developed a technique in the matter. Much of this cannot take effect until an adequate school plant is built. . . . The special units needed for the unique Mooseheart plan of education—shops, laboratories, libraries, studios, classrooms—are being included in the plan."

The aim of the Mooseheart educational system is to prepare the child for the responsibilities of citizenship; to make of each individual an economically efficient unit with proper social and political habits and attitudes. The girls enter directly upon an intensive vocational course at the completion of the junior high school period. They may take secretarial work, including bookkeeping, shorthand, typewriting, comptometry, filing, and dictaphone work, or commercial art, garment making, cafeteria management, institutional cooking, or prenursing. One year of accredited teacher training work is also offered. The work is intensive and constitutes direct preparation for employment.

The intensive vocational training for boys also begins after the ninth year of school. The vocation is selected upon the basis of individual interest in and capacity for the work. A minimum of nine and a maximum of fourteen hours per week is devoted to direct vocational training. In addition to the shop work, related drawing, mathematics, and science are required. There is a wide range of vocational offerings, including sheet metal work, machine shop, printing, office practice, painting, carpentry, building construction, electricity, and the like.

Contributions to Vocational Education from the Field of Education.—The modern scientific movement in education is a recent development. The widespread interest in professional improvement of teachers and increased efficiency of schools is a twentieth century product.

The Development of Natural Science After 1850.—The great strides in the physical sciences after 1850 were, perhaps, instrumental in creating an interest in education. Newton, Darwin, Huxley, Helmholtz, and others were leaders in the revision of the physical sciences. Experimentation in the physics of light and sound carried on by Benjamin Franklin and Count Rumford eventually led to research of a physiological character. Experimental psychology was the next step. During the early part of the nineteenth century there was a great increase in the amount of work done upon the eye and the ear.

The Beginnings of Modern Psychology.—Weber about 1825 made a number of important discoveries regarding the sense of touch and the perception of distance, temperature, and weight upon the skin. A whole movement grew up which used the same general method—the collection of a mass of data, the assimilation of the same, and a conclusion based upon the foregoing observations.

Professor Robert Woodworth in his book, "Dynamic Psychology" (page 8), notes the situation in 1870 as follows:

"We have the mental philosophers, best represented by Bain or the Herbartians, in Germany, disposed to devote their attention to the senses and intellect, the emotions and will, as matters deserving of study for their own sakes without regard to ulterior philosophical considerations; and on the other side we have a large and growing fund of information on the senses and sense perception, the speed of simple mental operations, and related topics, and we have a number of experimental procedures well worked out and known to be usable. The man in whom these two streams most definitely came together was Wundt."

Wundt, the Father of Modern Psychology.—Wundt worked in the field which he called "physiological psychology." In 1874 he published a book with that title. In 1879 he established at the University of Leipzig the first recognized psychological laboratory. Similar beginnings can be noted at Berlin,

Harvard, and Johns Hopkins universities by men who had studied under the physiologists.

The Work of Ebbinghaus.—Ebbinghaus is the next great psychologist to be considered. It was this man who demonstrated that experiment could be applied to memory. Many American psychologists, somewhat later, found practice and habit formation to be fruitful fields for experimental study.

The Development of Experimental Psychology.—The field of biological evolution was another scene of development, the influence of which was to be felt in education. In 1859 Darwin's "Origin of Species" appeared. The tremendous interest aroused in physical evolution by this volume spread to the sphere of mental evolution. Early works in this field were anecdotal and haphazard. In 1899 Edward Lee Thorndike pointed out the fallacy of this kind of evidence, and introduced the experimental study of animal intelligence.

Darwin himself made one of the first studies of mental development of children. It is to G. Stanley Hall, however, that credit is due for the accumulation of a great mass of observations of the biological development of children. Considerable use has been made of this material in the study of child psychology.

Galton's Work on Heredity.—Galton, a contemporary of Darwin, like Wundt can be considered a founder of modern psychology. He studied individual differences, mental traits, heredity and human variations. He sought to determine the relative importance of heredity and environment in the make up of human personality and behavior. Galton, according to Professor Woodworth ("Dynamic Psychology," page 12), "introduced the conception of mental tests, thus establishing connections between experimental psychology and the biological interest. In this line he was immediately followed by Cattell, and later by a host of psychologists, as the fruitfulness of this line of study has become evident."

William James and American Psychology.—No summary of the evolution of the modern trend in education is complete without a reference to the late Professor William James. Woodward ("Dynamic Psychology," pages 18–19) epitomizes the career of this great teacher most succinctly:

"Perhaps no one has better expressed in his writings the full scope and tendency of modern psychology than the late William James. He took as his background the older mental philosophy, especially of the English school, being however keenly aware of its shortcomings, and of certain necessary complements to be found in the mental philosophy of the Germans. Coming into psychology from the physiological laboratory, he retained his physiological point of view and was entirely hospitable to the new experimental psychology, and very early conducted experiments of his own. . . . Better than any other book, his great work on the "Principles of Psychology" can be taken as at once a summing up of the older psychology and an introduction to the modern point of view."

The Testing Movement in American Education.—The development of tests, both verbal and non-verbal is a study in itself. From the work done in France by Binet and Simon to the work done by Terman and others in America, the story is a long but progressive development. Trade tests, even in their crude form, have been of inestimable value in assisting directors of vocational education, while the verbal tests have indicated those students whose ability to do college work is questionable. Tests, however, are not the final word in the cataloging of an individual; neither is the I. Q. the sole basis of judging an individual's social worth. They are aids, nevertheless, and as such they have rendered great assistance to school administration and individual students.

The Results of the Testing Movement.—In the compendium, "Principles of Education," by Counts and Chapman, there appears on pages 568–569 a commonsense summary of the results of the testing movement in education:

"So productive has this movement been in its initial years that claims may be made that standardized mental and school tests, wisely employed, are capable of accomplishing the following ends:

"(a) Making definite the aims of education and determining the possibility, for individuals of varying mental levels, of attaining those aims.

"(b) Furthering the detailed analysis of the process of learning.

"(c) Aiding individual diagnosis and group analysis.

"(d) Improving classification and promotion, and educational and vocational guidance.

"(e) Improving school records and methods of reporting."

Trade Tests and Vocational Education.—The development of trade tests has yielded directly much of worth to vocational education directors; general intelligence tests have yielded as much indirectly. The realization that there are some for whom academic courses are ill-fitted and whose ability to pursue the same is questionable has made possible guidance into vocational channels many of those who, no doubt, would be deemed failures in the academic courses.

The Need for Better Aptitude-Discovering Devices in Vocational Education.—Mere failure in academic pursuits does not mean a guarantee of success in skilled industrial work. This fact bids fair to become universally recognized, and in its train will come the development of many and better aptitude-discovering and measuring devices. The work of Stenquist and Thurstone, to mention but two, has been of great significance; but the aptitude tests now available are too few in number and too crude to warrant their application to all children for prognosticating probable worth-while lines of occupational endeavor from among the myriad possibilities.

The Modern Teachers' Interest in Problems of Research. —When the success of experiments in applied psychology (and especially educational psychology) became evident, there was a rush by students of education to engage upon experimental problems. Individual differences, mental tests, methods and techniques, age-grade problems, retardation, and a host of kindred school topics were made the subjects of extended research. Professors in schools of education and teachers' colleges became focal points from which radiated countless students engaged upon research problems. Experimental classes and even schools were established. Public school systems invited scientific surveys, and individual classroom teachers many times changed

L. P. AYRES' STUDY OF THE "LAGGARDS" 185

their methods to keep abreast of the latest developments in scientific education.

The Work of Modern Educators, a Root of Vocational Education.—It is not the purpose of this study to set forth a full description of the work of the leaders in modern education, although such an enterprise would constitute a delightful and inspiring contribution to the annals of education. Vocational education owes a debt of gratitude, however, to great leaders in the field of education. Many practices in modern industrial education had their origin in the work and teaching of such men. The names of Inglis, Counts, Stanley Hall, Thorndike, Ayres, Strayer, and Van Denburg are only a few of those whose contributions, particularly in the field of school population, may be listed.

Retardation an Early Problem for Study.—One of the first subjects to be studied was retardation and elimination of pupils in the school system. In 1904, Superintendent Maxwell of New York City discovered the appalling amount of retardation in the schools of the city. He announced to a meeting of superintendents the fact that 39 per cent of the New York City children were above the normal age for their grades.

Leonard P. Ayres' Study of the "Laggards."—This statement stimulated systematic study of the problem of retardation. One of the most important contributions to this problem was the report of Leonard P. Ayres entitled "Laggards in Our Schools." This report showed that in thirty-one leading cities the average proportion of pupils who were a year or more behind the grade in which they should have been was 33.7 per cent. He reported that sixteen per cent of school children were repeaters. This extensive study listed among the causes for retardation late entrance, irregular attendance, and illness or physical defects. Mental inferiority was not listed as a cause in this study of 1909. At that period intelligence tests were in their infant stage, and Binet was just applying his work in a few institutions for the feeble minded.

Ayres listed among the administrative devices to reduce retardation the following:

1. Better compulsory attendance laws and better enforcement.
2. Better school census methods.
3. Medical inspection.
4. Better courses of study (to aim to fit the average child).
5. More flexible grading.
6. Better school records.

This monumental study had a wholesome effect upon the school world. Various cities of the more progressive sort gave much attention to the report, and sought to incorporate its suggested remedies in their schools.

Joseph Van Denburg's Study of Elimination.—A second important piece of research which had universal bearing upon American education was Joseph K. Van Denburg's study of the "Causes of the Elimination of Students in Public Secondary Schools of New York City" (Teachers College, Columbia University, Contribution to Education, No. 47, 1911).

In his introduction, Van Denburg states the problem tersely:

"With the growth of public high schools in the United States there has arisen a situation which, so far, has found no satisfactory explanation. In almost every city where high schools have been established, the entering classes have taxed the accommodations of the high school to the utmost. The number of pupils, however, who complete the course, is, when compared with the number which entered, so small as to excite surprise." (Page 1.)

The study indicated that approximately 10,000 pupils dropped out of the New York City high schools in 1906. One-third of these left before completing their first term, and more than half of all who left did not even complete the two terms of the first year of high school.

Van Denburg's study brought forth the significant fact that less than two per cent of the pupils that entered high school, during the period under study, reached the freshman class in college. In this study of elimination Van Denburg examined the grammar school graduates who entered high school. He surveyed their age, health, nationality, sex, occupation of parents, occupations of older sisters and brothers, financial status, vocational ambitions, and valuation of high school course.

One of his significant conclusions appears on page 78 of the study, as follows:

"For the most part then, at present, the high school curriculum, while open to many, is still planned largely to be of value to the extreme few. So we have, at least in the classical high school, an aristocratic institution of a very pronounced type under the guise of one supposed to be popular and democratic."

On page 188 there is found the germ of a later development:

"Among the many conclusions possible there seems at least one conclusion that we all must draw from this investigation taken as a whole; namely, that an extremely large percentage of the population enters the high school unwilling or unable to benefit properly by the instruction which is offered at present. To bar entirely those children from secondary education would be undemocratic and, in that it denies them equality of opportunity, unjust."

At this point Van Denburg suggests as a remedy the cosmopolitan high school with a variety of curricula offering a wide range of courses and giving ample opportunity for "try-outs."

The Studies of Thorndike and Strayer.—The work of Thorndike entitled "The Elimination of Pupils from School" (1907) is also significant. George Strayer is another who has done much in the study of age-grade problems. His work on "Age and Grade Census of Schools and Colleges" (Bureau of Education, Bulletin [1911] No. 5) has been of great value. As indicated in his tables (page 103), Strayer's investigation showed clearly that between 55 and 60 per cent of the pupils in the public elementary and secondary schools are of normal age, about one-third are below the grade where they might be expected to be according to their age, and less than 5 per cent are advanced beyond their age-group.

Thorndike found that for every 100 in school at 9 years of age, 9 leave at 12, 18 at 13, 23 at 14, 17 at 15, 14 at 16, 8 at 17. Most, he said, drop out from 13 to 15, feeling that school is not vital. (Thorndike, "Elimination of Pupils from School," pages 11, 47.)

George S. Count's Study of the Selective Character of American Secondary Education.—A remarkable survey was made in 1922 by George Sylvester Counts. The product of this research was the work, "The Selective Character of American Secondary Education," which was published by the University of Chicago in 1922. Counts stated his problem clearly on page 3 of the report:

"In view of the remarkable increase in high school enrolment and the changing concept of secondary education, it is becoming increasingly pertinent to inquire into the character of that student population which is attracted to the public high school." Again (page 4), "It is clear that a thorough study of the high school population is fundamental to the solution of all problems of organization and administration. The high school student should furnish the point of departure for the wise determination of high school policy and practice."

Counts studied the high school group in the cities of Seattle, Washington; St. Louis, Missouri; Bridgeport, Connecticut, and Mt. Vernon, New York. A total of some 17,992 cases was studied.

The character of the community was the first item surveyed in this study. Then after completing this preliminary investigation, Counts tabulated the parental occupation of the high school group. This vocational status he considered in conjunction with total enrolment, progress through the school, selection of the course of study, and expectations following graduation.

On page 87 of the study this summarization occurs:

"All the evidence presented indicates that the high school is, in the main, serving the occupational groups representative of the upper social strata."

The study continues with chapters on the public high school and the cultural level, the public high school and the immigrant, and the public high school and the negro. A later chapter is devoted to the psychological selection of the high school population.

In his "Conclusion and Interpretation" Counts states that the selective principle had not been abandoned in American secondary education. Also he notes the very close relationship between parental occupation and the privileges of secondary education. He shows that certain occupational groups were very well represented and others very poorly represented in proportion to their numbers in the general population. Among the former were found the five non-labor groups, with the professional service group first, followed by the proprietor group, commercial service group, managerial group and the clerical group. At the other end were found the groups of common labor, personal service, miners, lumber workers, and the like. Those classes which are least well represented in the last year of the high school, the report showed, were even less well represented in college.

The importance of the family and its place on the social ladder as a powerful factor determining school attendance was indicated by the evidence. According to the data gathered Counts reported that the age-old adage of "to them that hath shall be given" is as true today of American secondary education as it has been through the centuries. Thus secondary education remains largely a matter for family initiative and concern, and reflects the inequalities of family means and ambitions.

These Studies Have Constituted Roots of Vocational Education.—And now what has been the significance of these studies to general education and to vocational education? These studies (and many others which cannot be mentioned within the limits of a work of this nature) indicated that the academic high school was failing to meet the needs of a great proportion of the population. The realization came, after years of waiting, hoping, and fighting, that the college preparatory course should not be considered the *raison d'être* of the high school. The retardation, elimination, and selective character studies brought before educators the problem of reorganizing the schools so that they might render more service to the contemporary world. There followed extensive studies in curriculum development and enrichment. The work of Judd, Bobbitt, C. O. Davis, and Philip Cox was called forth by these conditions. At the same time there came the present developments in vocational education.

A Recent Study of the School Population.—A very recent study that bears upon the point has been made by the "Committee on Vocational and Industrial Education of the High School Teachers' Association (of New York City)." The report of the findings of this group is summarized in the columns of "The New York Herald Tribune" of December 23, 1929. The headline blazes forth with the statement that the high school population is 37 per cent below normal.

The report states in part:

"In the twelve Brooklyn high schools, where 6,278 pupils are enrolled in the fourth term, 3,905 students have been promoted to the fourth term progress grade after three terms of work, while 2,273 have taken four, five, six and some even seven terms to reach the fourth term. That is, 36 per cent of all the high school pupils in Brooklyn have not been able to reach the fourth term after three terms of work.

"In the seven high schools of Manhattan, where 3,616 are enrolled in the fourth term, 2,443 students have been promoted to the fourth term after three terms of work, while 1,173 have taken four, five, six, and seven terms to reach the fourth term. That is, 32 per cent of all the fourth term high school students in Manhattan have not been able to reach the fourth term after three terms of work.

"Similar surveys were made in the Bronx, Staten Island, and Queens. The per cents below grade in these boroughs were found to be 32, 40 and 46 per cent respectively.

"We hope to investigate further the work of pupils in the vocational and industrial courses to show that, when certain pupils are allowed to take these courses, their abilities have their proper fulfilment and that they do excellent work." (Statement of Mrs. Dora Thompson, Chairman of the Committee.)

A further statement attributed by the "Herald Tribune" to the group chairman was that the hope of the committee was to bring to light more facts by which to impress the Board of Superintendents with the necessity for increasing the amount of vocational and industrial training in the schools.

THE WINNETKA AND DALTON PLANS

The Winnetka and Dalton Plans.—The recognition of individual differences has greatly influenced the character of modern education. From the earliest studies of Galton down to the present there has been much work done in the field of individual differences. The names of Hall, Thorndike, Strayer, and Whitley will ever be associated with this phase of experimental education.

Practical applications of the recognition of individual differences have found expression in the Winnetka Plan and the Dalton Plan of education.

The Winnetka Plan has been in operation in the city school system of Winnetka, Illinois, for several years. Mr. Carleton Washburne, the superintendent of schools, attributes the development of this plan to a thoughtful solution of a problem with which he was confronted in his early days of teaching. When Washburne began his work of teaching at La Puente, California, he found his classes composed largely of Mexican children of greatly varying levels of achievement. He devised a plan of individual teaching whereby each pupil could proceed at his maximum efficiency. His success was repeated the next year at Tulare, in the San Joaquin Valley, with a class of misfit pupils. During the next five years Washburne worked with Dr. Frederic Burk at the State Normal School at San Francisco on the problem of individual differences. In 1919 he was called to the position he has since held in Winnetka.

Martin Stormzand in "Progressive Methods of Teaching," page 363, lists the steps in the Winnetka Plan:

"(1) Clear, definite goals must be established;
(2) There must be complete diagnostic tests prepared;
(3) Self-instructive and self-corrective practice materials must be provided."

A somewhat similar plan of individual instruction is that known as the Dalton Plan. This was first worked out by Miss Helen Parkhurst for use in a private school at Dalton, Massachusetts. Since then, it has been adopted by the public school system of that city.

The essential characteristics of this plan include (1) departmentalized teaching in all grades; (2) self-directive assignments or "job contracts"; (3) a total absence of any class schedule; (4) individually directed study with as much help from the teacher as the pupil asks for; (5) individual progress of the pupil in the various subjects and through the various grades; (6) a graphic record kept by the pupil of his own progress.

This movement toward individual instruction is psychologically sound, and its shortcomings, other than a possible loss of the values of group activity, may be classed as administrative rather than pedagogical. Both the Dalton Plan and the Winnetka Plan have left their stamp on industrial education. The individual job, the job sheet, and the various systems of individual progress are coincident points of contact. In this connection, Robert W. Selvidge's book, "Individual Instruction Sheets—How to Write and How to Use Them," is a significant contribution.

Job Analysis and Job Sheets.—With the great movement toward increased division of labor and greater specialization of processes in twentieth century industry has come a greater interest in the various jobs performed by industrial workers. Efficiency in industry has meant that each job must be performed as perfectly as conditions permit. The job, then, has been taken as the unit upon which the whole production flow depends. There has grown up a movement to study the job in great detail, its requirements, its component parts, and its relation to other jobs in the shop.

This movement has been inclusive and has concerned itself with incidental occupations as well as with the study of all the jobs within a given business or industrial organization. Job analyses have been made for the job of works manager of a factory as well as for the job of green-sand moulder in the foundry of the same plant.

Engineering companies were among the first groups to become aware of the serious shortage of men capable of filling executive positions. Graduates of engineering colleges for the most part appeared to lack certain training necessary for the handling of men, materials, and money. Business and commercial houses soon agreed that there existed a dearth of embryonic leaders. The result has been the establishment of curricula of "Industrial (or Administrative) Engineering" in the schools of engineering, and "Business Administration" in schools of commerce, accounts, and finance. Already the ideals and programs of these courses are well known and their graduates are expected to fill the ranks for which those curricula train.

In general business and industry seek to replenish their ranks with capably trained workers. In order to create proper courses of study and training programs extensive surveys of real conditions in industry are necessary. Industrial and commercial leaders were not slow to realize that the distinctive features of each job constituted the foundation upon which job training programs were to rest. Specific job requirements had to be met. Job analysis was the method devised to determine these requirements.

Job analysis is described as "the search for habits necessarily used by a workman on the job" by Strong and Uhrbrook in their book "Job Analysis and The Curriculum," page 22.

Job analysis consists in the listing of the various skills and the related technical knowledge which is required in the performance of any one job in an occupation. Occupational analysis consists in breaking the occupation up into a number of subdivisions and in securing an inventory of the entire mass of related technical knowledge which is involved in the practice of the occupation. In the making of occupational analyses two procedures have been followed. According to the first procedure the skilled content of the occupation is broken down into a number of fundamental operations and technical content is identified as it applies to each operation. According to the second method, the occupation is broken down into jobs as recognized by practitioners of the occupation, these jobs being used in a similar way to identify related content. Such inventories of content and skills, whether secured by either procedure, form a basis for

194   THE CONTRIBUTION OF SOCIAL AGENCIES

the preparation of instruction sheets, but do not in themselves constitute either courses of study or instruction sheets.

In the actual preparation of instruction sheets the device designed to present to the student the necessary instructions, information, and directions for the successful performance of a task, Robert W. Selvidge's book "Individual Instruction Sheets" (The Manual Arts Press, 1926) is a competent guide.

Fred W. Taylor, in his volume "Shop Management," page 83, published in 1911, wrote: "Each job should be carefully subdivided into its elementary operations, and each of these units should receive the most thorough time study." Taylor's chief interest was time and motion study; nevertheless, it is to Taylor that the movement for job analysis owes its early impetus.

In 1920 the Carnegie Institute of Technology started surveys in three distinct lines, building construction, commercial printing and the metal working industries, in order to determine the job requirements of executive positions in these occupations.

The complete reorganization of industrial plants in harmony with the principles of scientific management also contributed to the job analysis movement. The almost universal establishment of the centralized employment office in industrial plants created a definite need for detailed information concerning each job to be filled.

The technique of modern job analysis ranges from the mere listing of the tools used and operations performed on the job to the detailed measurement of the job made by scientific tests. The camera and motion picture machine have also been used extensively to give the actual record of a man's physical movements as he performs the job.

The advent of the trained employment manager and industrial psychologist was followed by the appearance of the job analyst who is capable of making job surveys.

Among the uses to which job analyses may be put, there could be listed:

1. Explanations to applicants of job requirements.
2. Determination of physical, mental and educational requirements of the job.
3. Relation of age to efficiency.
4. Determination of the amount and kind of previous training required.
5. Determination of the time required for the efficient performance of the job.
6. Determination of what jobs might be filled efficiently by women.
7. Reduction of labor turnover.
8. Development of training programs.

The work of the Bureau of Personnel Research, the Cleveland Educational Survey, the Federal Board for Vocational Education, the United States Department of Labor (Descriptions of Occupations), and the Division of Vocational Education of the University of California may be mentioned as significant contributions to this work. The reports of these agencies have had wide circulation, but the publications of the last named, the Division of Vocational Education, University of California, are especially inclusive and intensive studies. The titles of the monographs in the appended list are indicative of the scope of the work.

LIST OF PUBLICATIONS OF THE DIVISION OF VOCATIONAL EDUCATION, UNIVERSITY OF CALIFORNIA

Trade and Industrial Series

No. 1. Bulletin No. 12. Analysis of the House Carpenter's Trade. March, 1923.
No. 2. Bulletin No. 13. Analysis of the Cabinetmaker's Trade. September, 1923.
No. 3. Bulletin No. 15. Analysis of the Plasterer's Trade. April, 1924.

Agricultural Education Series

No. 1. Bulletin No. 8. Job Analysis Applied to the Teaching of Vocational Agriculture. May, 1922.

No. 2. Bulletin No. 11. Farm Mechanics for California Schools. November, 1922. (Out of print.)
Agricultural News Letter. (Monthly.)

Part-Time Education Series

Division Bulletin No. 1. Syllabus of an Introductory Course on Part-Time Education. January, 1920. (Out of print).
No. 1. Leaflet No. 1. A First Reading List for Administrators and Teachers in Part-Time Schools. August, 1920. (Out of print).
No. 2. Leaflet No. 2. The Work of Coordination in Part-time Education. November, 1920. (Out of print).
No. 3. Bulletin No. 2. An Analysis of Department Store Occupations for Juniors. December, 1920.
No. 4. Bulletin No. 3. Coordination in Part-time Education. March, 1921. (A revision of Leaflet No. 2).
No. 5. Bulletin No. 4. An Analysis of the Work of Juniors in Banks. May, 1921.
No. 6. Bulletin No. 5. An Analysis of Clerical Positions for Juniors in Railway Transportation. August, 1921.
No. 7. Leaflet No. 3. Selected Reading List for Administrators and Teachers in Part-time Schools. September, 1921.
No. 8. Bulletin No. 6. Part-time Continuation Schools Abroad—Reprints. November, 1921.
No. 9. Bulletin No. 7. The Work of Juniors in the Telegraph Service. April, 1922.
No. 10. Leaflet No. 4. Recreational Reading for Part-time and Continuation Schools. March, 1922.
No. 11. Bulletin No. 9. The Work of Juniors in Retail Grocery Stores. July, 1922.
No. 12. Bulletin No. 10. Third Annual Report of the Director of Part-time Education. Stockton, Calif., October, 1922.
No. 13. Bulletin No. 14. The Administration of the Part-time School in the Small Community, Part One. April, 1924.

News Notes

Part-time News Notes, Vol. I, Nos. 1–8. November, 1920–May, 1922. (Nos. 1–5, 7 out of print).
Vocational Education News Notes, Vol. II, Nos. 1–7. September, 1922–April, 1924. (Nos. 1 and 5 out of print).

Among the many individuals whose contributions have been significant in this field particular mention can be made of Charles R. Allen, Clyde A. Bowman, Frank Galbraith, Layton S. Hawkins, R. R. Lutz, Robert W. Selvidge and Frank Cushman.

FEDERAL AID TO VOCATIONAL EDUCATION

The Federal Government Has Often Aided Vocational Education.—Prior to 1917 the Federal Government had given assistance on several occasions to various forms of vocational education. Some references have been made already to these grants, but further discussion is necessary. The Morrill Act of 1862 provided grants of public land to the several states and territories for the establishment of colleges of agriculture and mechanic arts. The Hatch Act of 1887, the next instance of governmental aid, provided for an annual appropriation of $15,000 from the proceeds of the sale of public land for the maintenance of agricultural stations of research and experimentation. The second Morrill Act (1890) provided additional funds for land grant colleges.

The Smith-Lever Act, 1914.—Finally the Smith-Lever Act of 1914 provided wisely and liberally through national grants to the states for agricultural education for the mature farmer through farm demonstration and farm-extension work. The provisions failed to include grants for teachers, supervisors, and directors of agricultural subjects in district agricultural high schools or in agricultural departments of rural high schools. The grants were divided among the states according to their rural population. The act further required that each state, or the local authorities in it, appropriate an equal amount to be used in accomplishing the work provided for by the act.

The Demand for Federal Aid for Endeavors of Less Than College Grade.—All these grants had been for the most part for vocational education of college grade. ''Meanwhile there had arisen a widespread demand for Federal aid for vocational education in the lower schools. Among its advocates were educators, reformers, manufacturers, and labor organizations. The American Federation of Labor, from 1903 on, consistently and unremittingly advocated the establishment of industrial education in the schools. A very large number of labor unions went on record from time to time as approving bills for vocational education that happened to be before Congress. The National Society for the Promotion of Industrial Education, including among its members representatives of practically all the trade unions and of all the prominent manufacturers associations, was formed in 1906. This society played a prominent part in the movement. The Chamber of Commerce of the United States in 1913 and again in 1916 adopted resolutions strongly endorsing the principle of liberal appropriation by the Federal Government for the promotion of vocational education in the states." ("The Federal Board for Vocational Education. Its History, Activities, and Organization." W. Stull Holt, page 3.)

Congress Appointed the Commission on National Aid to Vocational Education in 1914.—Pressure of this nature called forth much comment and discussion. There was not a session of Congress from 1910 until 1914, when one or more bills dealing with some phase of vocational education were not introduced. The culmination was reached in the creation of the Federal Commission on National Aid to Vocational Education in June 1914. Congress authorized the President to appoint a group of nine men who were to consider the subject of national aid for vocational education and report their findings.

The Report of the Congressional Commission.—The commission held extensive hearings and presented a lengthy report. Its report is briefly summarized in the service monograph (number 6) of the Federal Board for Vocational Education, page 4.

"There is a great and crying need of providing vocational education of this character for every part of the United States—to conserve and develop our resources; to promote a more productive and prosperous agriculture; to prevent the waste of human labor, to supplement apprenticeship; to increase the wage earning power of our productive workers; to meet the increasing demand for trained workers; to offset the increased cost of living. Vocational education is, therefore, needed as a wise business investment for this nation, because our national prosperity and happiness are at stake, and our position in the markets of the world cannot otherwise be maintained."

Prosser and Allen Summarize the Work of the Commission.—Prosser and Allen in their book, "Vocational Education in a Democracy" (page 424) commenting upon this statement of the Federal Commission state: "The two great assets of a nation which enter into the production of wealth, whether agricultural or industrial, are natural resources and human labor. The conservation and full utilization of both of these depend upon vocational training."

The Need for Federal Aid Epitomized.—The same authors find that the social and educational need for vocational education is just as urgent as the economic need expressed by the Federal Commission. On page 424, there appears the following:

"It [vocational education] is needed to democratize the education of the country:

"a. By recognizing different tastes and abilities and by giving equal opportunity for all to prepare for their life work;

"b. By extending education through part-time and evening instruction to those who must go to work in the shop or the farm.

"Vocational education is also needed for its indirect but positive effect on the aims and methods of general education:

"a. By developing a better teaching process through which children who do not respond to book instruction alone may be reached and educated through learning by doing;

"b. By introducing into our educational system the aim of utility to take its place in dignity by the side of culture and to connect education with life by making it purposeful and useful."

The Vocational Education Act of 1917. (The Smith-Hughes Act).—The report of the commission led to the passage of the Vocational Education Act of 1917 (popularly known as the Smith-Hughes Act). Under the provisions of this monumental beacon on the path of progress the states got three appropriations on condition that they or the local communities or both would spend an equal amount for the same purposes.

The first was for the purpose of cooperating with the states in paying the salaries of teachers, supervisors, or directors of agricultural subjects. This fund began with a $500,000 nucleus and was to increase annually until 1926 when it amounted to an annual item of $3,000,000. This sum was divided among the states according to their rural population. The second appropriation was for the purpose of assisting the states in paying the salaries of teachers of trade, home economics, and industrial subjects. It also started with the same sum and advanced toward the same maximum as the first fund. The distribution of the money for this appropriation was on the basis of the state's urban population. The third grant went for aid in preparing teachers of trade, industrial, agricultural, and home economic subjects. It began with $500,000 and reached its maximum of $1,000,000 in 1921. This sum was allotted upon the basis of total population.

The act also provided that in order to secure Federal aid the state had to accept the provisions of the act and create or designate a state board of at least three members having the power to cooperate with the Federal Board.

The Creation of the Federal Board for Vocational Education.—It further provided for the creation of the Federal Board for Vocational Education to consist of the Secretaries of Agriculture, Commerce, and Labor, the United States Commissioner of Education, and three citizens of the United States to be appointed by the President, by and with the advice and consent of the Senate. Of these three citizens one was to be a representative of manufacturing and commercial interests, the second of agriculture, and the third of labor.

The act states that the board shall have the power to cooperate with the state boards in carrying out the provisions of the act and that it shall be the duty of the board to make or cause to be made, studies, investigations and reports with particular reference to their use in aiding the states in the establishment of vocational schools and classes and in giving instruction in agriculture, trades, industries, commercial pursuits, and home economics.

The board upon its creation was quick to act with the states. By December 31, 1917, every one of the forty-eight states had accepted its provisions either through the legislature or governor. Each one had submitted plans for the current year which had met with the approval of the board. Stull Holt, page 7, says: "The total number of pupils enrolled in vocational courses in schools federally aided for the year ended June 30, 1918, was 164,186. For the year ended June 30, 1921, the number had increased to 305,224."

The "Thirteenth Annual Report to Congress of the Federal Board for Vocational Education," (1929), contains some significant figures.

The number and sex of pupils enrolled in vocational courses approved by the Federal Board for 1929 as given in table three, page 60, is:

```
Both sexes ............................... 1,047,957
Male ..................................... 591,658
Female ................................... 456,299
```

The number and sex of teachers of vocational courses in specified types of vocational schools federally aided is given in table two, page 58, of the same report:

```
Total .................................... 22,144
Male ..................................... 15,299
Female ................................... 6,845
```

The Federal Board for Vocational Education is the designated governmental agency responsible for the promotion of two types of social service. It is charged with the responsibility of administering the national vocational education act of 1917 and the civilian vocational rehabilitation act of (1920).

Dean Russell on Vocational Education.—Frank Leavitt in his "Examples of Industrial Education" (page 40) quotes Dean Russell of Teachers College, Columbia University, as saying:

"It is the boast, too, of most Americans that our great public school system provides alike for every boy and girl taking advantage of it. This is half true and dangerous, as half truths are. The fact is, the American system of education grants equality of opportunity to those who can go to college and the university. It takes little account of the boy, and still less of the girl, who cannot have, or does not wish for, a higher education. Ten millions of those now in our elementary schools, who will be compelled to drop out to earn a livelihood, will have missed their opportunity. But why? Do we in America have need only of professional men and men of affairs? Are those who pay the taxes and do the rougher work of life to be denied the opportunity for self-improvement? Are only those who can afford to stay in school to reap the advantages of an education? In a word, what are we doing to help the average man better to do his life work and better to realize the wealth of his inheritance as an American citizen? The questions raise the problem of vocational training for those who must begin early to earn their living. It is, in my judgment, the greatest problem of the future, and one which we may not longer disregard and yet maintain our standing as a nation."

Snedden on Vocational Education.—Professor David Snedden of Teachers College, Columbia University, in an address delivered before the National Congress of Parents and Teachers on December 6, 1929, said, according to quotation in the "New York Herald-Tribune," December 7, 1929:

"Many leading educators are still so uninformed as to real educational values that they still think that good general education, as personified in the four-year high school curricula, makes an important contribution to vocational competency. Educators should expect the large majority of students to stay in schools of general education until eighteen years of age.

"Educators should, on the other hand, see that for nearly all trade, factory, commercial, and even agricultural vocations from one to a dozen vocational schools for each distinctive line is all that will be required, as is so clearly the case now with schools of medicine, engineering, teaching, and agricultural leadership."

Dr. Snedden in the same address also said that if vocational education in public schools is to be made useful it must be as highly specialized as the training in the learned professions. Carpentry, brick laying, and other manual occupations should be taught in as specialized a manner as medicine, law, and theology are now taught.

These are only two famous educators, yet their remarks are significant as examples of current thought along the line of vocational training. Even though many may disagree with the ideas, methods, and procedures of these and other educators, the important thing is that vocational education has had its prophets and workers. The whole movement has been stimulated by such disagreements and discussions.[2]

Social Workers and Vocational Training.—Social workers, acting through philanthropic and charitable institutions, have exercised much influence in aiding vocational education. The task of the social worker has usually been prompted by the desire to ameliorate the hard and difficult conditions of the unfortunate. Consequently those workers have been among the first to make real studies of actual conditions under which industrial workers live. In home and shop visits, the social workers have collected valuable data and statistics. Immigration, school attendance, wage scales, hours of labor and the like are included in the reports of their findings. Their work in prisons and reformatories has usually been characterized as helpful.

Industrial training has been instituted in prisons, and Leavitt in his "Examples of Industrial Education" mentions several such examples and quotes from the reports of the National Prison Association.

The Philadelphia House of Refuge has trade schools, founded, according to Leavitt, so that boys may be trained for useful and skilful pursuits. It assists them to find employment at satisfactory wages when they leave the House.

The North Bennett Street Industrial School (of Boston) has conducted industrial classes for many years. It was a pioneer in demonstrating that the last two years of elementary school could be made vocationally valuable to boys and girls who enter industrial occupations at an early age.

---

[2] Dr. Struck's book, "Foundations of Industrial Education," gives a splendid treatment of the personalities that have actively engaged in the movement for vocational education.

Scientific Societies Have Been Part of the Movement for Vocational Education.—The various scientific societies have studied the problem of industrial and trade training, and the contributions made by the researches which they have conducted have been significant.

The American Society of Mechanical Engineers has maintained a committee on "Education and Training for the Industries." This committee has for its purpose the study of industrial education. Hardly an issue of the society's magazine, "Mechanical Engineering," is published without a paper by this committee. The reports of this committee form an integral part of the conventions of the society.

To indicate the scope of the work of the American Society of Mechanical Engineers, acting through the committee on "Education and Training for the Industries" a few studies that have been published might be mentioned. "Extension and Correspondence Schools" by James Mayer (Mechanical Engineering, Vol. 45, No. 3); "Schools for Apprentices and Shop Training" by R. L. Sackett (ibid); "Industrial Education as Represented in Schools" by C. R. Richards (ibid); "Industry's Interest in Industrial Training" by Magnus Alexander (Mechanical Engineering, Vol. 47, No. 2); "Has the Need for Apprenticeship Passed?" by W. A. Viall (Mechanical Engineering, Vol. 48, No. 5), "Technical Training in Industry" by John Kottcamp (Mechanical Engineering, Vol. 47, No. 8). This list indicates, by a mere selection of a very few of the articles, the interest displayed in trade training by a group of organized engineers.

The Society for the Promotion of Engineering Education is another group interested in the development of technical training. This group publishes the magazine, "The Journal of Engineering Education," which is often replete with articles and studies pertaining to vocational education.

Many Publications Have Advocated Vocational Education.—"The Industrial Arts Magazine," "Industrial Education," "Junior-Senior Clearing House," "Journal of the American Vocational Association," "Bruce's Specifications Annual," the annual and monthly reports of the Federal Board for Vocational Education and of the various state and local boards, particularly New York, California, Wisconsin, Pennsylvania, New Jersey, and Essex County, New Jersey, are publications whose influence on vocational education is not to be overlooked.

The Weaknesses of the Traditional Education Become More Evident with the Advent of Scientific Study.—These social groups have noted various weaknesses in the existing social structure, and more especially the educational field, and have sought to remedy these shortcomings. It is questionable how much individual efforts can do, but it is almost certain that individual efforts, if worthy and sincere, lead to more concerted action. Therein lies a large measure of the contributions of the smaller groups. They experiment and point the way toward progress.

Manual Training Rested Upon the Faulty Foundation of the Transfer of Training.—The realization that manual training as a subject embodying functioning transfer was psychologically false is among the most poignant of the conditions contributing to the vocational education movement. Modern psychology and scientific research sounded the death-knell of the so-called Russian system of manual training.

The Changed Character of the Secondary School Population, a Root of Vocational Education.—A second important contribution to vocational education was the realization that the secondary school population had changed in character as well as in numbers. The growth of legislation for compulsory school attendance and the stricter enforcement of existing attendance laws, together with the public's greater valuation of "an education" and greater economic independence, brought to the school a wider cross-section of human society. The attention of the school was focused on that large group which would relatively early find places in vocational pursuits. Provision for the education of these was a necessary result.

Changed Home Conditions a Root of Vocational Education.—The influence of the Industrial Revolution in transferring the center of production from the home to the shop revolutionized American home life. The consequent urbanization of population reduced the social and educative value of domestic chores and like activities. Actual vocational activities grew up so that the child could be of service to his family and to society.

The decay of the apprenticeship system had reached its culmination by the beginning of the twentieth century. Entrance to journeymanhood or mastership was restricted. The school was looked to for relief.

Immigration restriction and a host of other causes reduced the army of skilled workers. The school again was expected to furnish capable workers for industry.

The Growth of Democracy, a Characteristic of Modern American Life.—Another contributory factor is one of a social nature which is difficult to classify or measure. It is the growth of a scientific spirit, a willingness to face the facts of life. The result has been the development of a liberal and democratic spirit. Evidences of this spirit have pervaded the realms of industry, morals, and education. The industrial education implications reside in the more general valuation of education in terms of conduct and action rather than content and method.

Adult Education a Mass Movement in the Twentieth Century.—Perhaps at no time since the thirteenth century has the desire for education so nearly approached a mass movement. The entire field of adult education has been opened up, and many leaders in current thought point to education as a guarantor of a new and better civilization.

Joseph Hart in the introduction to his "Adult Education" says, "Education must mean more than schooling for the youth of the country."

He stresses the need for adult education in a democracy and looks for a release of much latent power through individual development. He notes that education lags behind other developments and quotes copiously from educators, from Rousseau and the "Naturalistic" school to the present time with the "new school movement" to which John Dewey has con-

208 FEDERAL AID TO VOCATIONAL EDUCATION

tributed so lavishly. The influence of experimental schools has eventually found expression in the public schools and in such movements as have made possible Gary and Winnetka.

Educators Whose Contributions to Vocational Education Are Significant.—In much the same way that dates and eras will be listed as milestones in "Part Seven" of this work, so can the names of certain educators be listed whose contributions, whether direct or indirect, have been of great value to the furtherance of vocational education. This chapter, necessarily brief, can include no more than a listing of the names of a few of those of this era who deserve to be mentioned along with the others whose work has been touched upon in this section. A complete list is beyond the scope of this work.

Charles R. Allen—Educational Consultant Federal Board for Vocational Education.
Frederick J. Allen—Vocational guidance.
Richard D. Allen—Vocational guidance.
Leonard P. Ayres—Research in the field of the school population.
Earl W. Barnhart—Chief of the Commercial Education Service of the Federal Board for Vocational Education.
Adelaide Baylor—Chief of the Home Economics Education Service of the Federal Board for Vocational Education.
Robert Beebe—Director of Vocational Education in Essex County, N. J. Authority on county curriculum reorganization.
Charles A. Bennett—Editor and Author Historical Aspects of Trade and Industrial Education.
Verne A. Bird—Director of Vocational Education at Mooseheart, Ill.
Meyer Bloomfield—Director of Vocational Bureau, Boston, Mass. Guidance and placement.
J. Franklin Bobbitt—Curriculum reorganization.
Frederick G. Bonser—Industrial Arts.
John Brewer—Authority in the field of vocational guidance.
Frank Bruce—Publisher in the field of industrial arts and vocational education.
L. H. Carris—Former director of the Federal Board for Vocational Education.
W. W. Charters—Director of Research, Ohio State University.
Robert L. Cooley—Director of the Milwaukee Vocational School.

CONTRIBUTIONS ARE SIGNIFICANT

George S. Counts—Experimenter and research worker in the field of the selective character of secondary education.

Philip W. L. Cox—Curricula reorganization and socialized activities in the schools.

Frank Cushman—Chief, Trade and Industrial Service Federal Board for Vocational Education. Foremanship training.

Arthur Dean—Author and lecturer.

Theodore H. Eaton—Professor of Rural Education, Cornell University. Studies on education and vocations.

A. H. Edgerton—Professor of Industrial Education, University of Wisconsin. Industrial arts.

John F. Friese—Director of Manual Arts and Vocational Education, St. Cloud, Minn. Industrial arts and evening school education.

Frank Galbraith—Industrial engineer. Time and motion study.

Layton S. Hawkins—Former assistant director of the Federal Board for Vocational Education. Analyses and individual instruction sheets.

Franklin J. Keller—Authority on continuation schools and vocational guidance.

Ronald W. Kent—Assistant director of Essex County (N. J.) Vocational Schools. Expert on county curriculum reorganization.

H. D. Kitson—Vocational guidance.

Frank Leavitt—Study of "Examples of Industrial Education."

Edwin A. Lee—Director of the Division of Vocational Education, University of California.

Robert J. Leonard—Studies of professional education. (Deceased.)

Henry C. Link—Employment psychology. Job specifications.

Paul S. Lomax—Professor of Commercial Education, New York University. Authority on commercial education.

Arthur B. Mays—Author, industrial education.

John Van L. Morris—Employee training.

Ralph E. Pickett—Professor of Vocational Education, New York University. Authority on vocational and industrial arts education.

Charles A. Prosser—Director of Dunwoody Institute, Minneapolis, Minn. First Director of the Federal Board for Vocational Education. Author and pioneer.

G. A. Schmidt—Professor of Agricultural Education, Colorado Agricultural College.

Herman Schneider—Pioneer in cooperative education.

Robert W. Selvidge—Job sheets, individual instruction sheets, job analysis.
Z. M. Smith—State director of vocational education, Indiana.
David Snedden—Pioneer and philosopher in vocational education.
John L. Stenquist—Tests and measurements in mechanical aptitude.
G. D. Strayer—Studies in school population and school administration.
L. L. Thurston—Tests and measurements in education.
S. Vaughan—Author (with Mays) of "Content and Methods of Industrial Arts."
Garton K. Wagar—Director of Vocational Education, New Rochelle, N. Y. Authority on industrial arts education.
Lewis A. Wilson—Promotional work in state vocational reorganization.
J. C. Wright—Director of the Federal Board for Vocational Education. Co-author with C. R. Allen of books on supervision and administration of vocational education, efficiency in education and efficiency in vocational education.

Part Four

ORGANIZED BUSINESS AND VOCATIONAL EDUCATION

This section deals with the attempts of organized business to train its employees for more efficient participation in the activities of the shop. From the days of the guilds down to the present, the wise employer has sought to improve the services rendered by his workers. The "factory" school in one form or another has occupied a prominent place in the world of vocational education.

In the preparation of this section of the work the brochures and catalogs of a few representative company schools have been studied. Also, a questionnaire was addressed to fifteen representative industrial and commercial firms asking merely when and why they instituted their educational departments. In most cases they either could not or did not choose to answer the question why.

ORGANIZED BUSINESS AND VOCATIONAL EDUCATION

Modern Industry a Recent Evolution.—A survey of industrial changes during the past fifty years has revealed countless new methods, techniques, and processes in the fabrication and manufacture of articles. New types of machinery, improved jigs and fixtures, and modern tools have startled the most imaginative specialists in the industrial world.

"Big Business" a Twentieth Century Product.—No less striking has been the change in business during the same period. "Big business" has evolved. Today the gigantic corporation has proved itself an efficient unit in modern society. The control of the manufactured product is the work of large business interests. The financing, advertising, and selling of an article are taken care of today by huge corporations, the like of which were unknown fifty years ago.

The Scientific Attitude Characterizes Many Business Executives.—Even a cursory study of these business enterprises shows that they are controlled by men who, for the most part, are ready to investigate and try the plans and schemes of research offered them by specialists. The success of many is due to this characteristic daring and adventuresome spirit, backed by the findings of research specialists.

The Results of Scientific Investigations Have Been Incorporated in the Modern Business World.—Modern theories and doctrines of finance, organization, and psychology have found place in the business world. The human and business engineer has set out to make business as efficient as the manufacturer and mechanical engineer have made industry. The various bureaus of business research, the Taylor Society, and the management engineering groups are living witnesses to the fact.

Human Engineering Has Developed.—The personnel movement in business and the emphasis placed on human relationships are of recent date. The realization that human machines are as important as steel ones was slow in coming, but it is here now. With this has come interest in education, and more especially vocational education. Corporations have established schools of their own to train workers. Such a departure, resulting as it does in the expenditure of dollars and cents, demands a searching study. Why has business undertaken the work of teaching its men and women? A survey of the more prominent corporations' schools has revealed interesting answers to that question.

A. The Boston Elevated Railway Company.—The Boston Elevated Railway offers an interesting educational program. The management of the corporation instituted this program in cooperation with the Division of University Extension, Massachusetts Department of Education, in the belief that some employees would like to have instruction of a practical nature along with their every-day work. The courses conducted by the division proved popular, and led to the development of a more ambitious program. It is difficult to say just what really started the movement. The present director and educational advisor, Henry H. Norris, states that the origin was simply in line with a general educational movement throughout the industry. He also states that Mr. Edward Dana, general manager of the Boston Elevated Railway, has been for a number of years chairman of the committee on education of the American Electric Railway Association. He infers that there is some connection between this fact and the other fact that he had been interested in developing the present educational program of the company.

The plan started with the idea of giving employees such instruction as they desired and not forcing anything on them. The management feels that by broadening his outlook, the employee becomes more efficient and happier in his work. Mr.

Dana each year appoints a committee on education, representing the several departments, and this committee is charged with the responsibility of recommending a program for the ensuing season. Practically all the instruction is on the employees' own time.

In the official organ of the American Electric Railway Association, January, 1926, there is an article by Henry Norris entitled "What One Railway is Doing Along Educational Lines." The following statements indicate significant ideas on educational policy.

"As electric railway employees are normal human beings, it is to be presumed that a reasonable proportion of them are capable of self-improvement and eager for it. Many of them continue the studies begun in school, with the aid of correspondence courses, night schools, etc. Knowledge of such efforts at self-improvement is helpful to employers in permitting working persons to be promoted as opportunity offers. It is obviously a paying proposition for employers to encourage a continuation of study in every practical way.

"Electric railways, the railroads, other public utilities, and manufacturers, in fact employers generally, are coming to realize that they have an educational function to perform with profit to themselves and their employees. This function is to supplement the work of the school, but not to duplicate it. It includes coordinating the local educational facilities for the benefit of employees, and supplementing these with such instruction as cannot be obtained conveniently elsewhere.

"The increasing use of expert educational service by employers is an encouraging sign of the times. Even closer cooperation between industries and educators is highly desirable.

"Last year (1925) the educational program took the form of five series of conferences, held respectively for employees of the transportation, stock, and shop departments, and for the women employees. These series consisted each of from ten to twenty group conferences on topics carefully selected for the season in advance. This group conference plan was highly successful, enlisting the cooperation of nearly 1,100 employees."

Now courses are held in over twenty groups including public speaking, accident prevention, correspondence and reports, signals, business English, psychology, current events, automotive maintenance, and many others.

Three types of methods are employed according to the subject and the class group. They are (1) the straight conference method, (2) the group conference plan, and (3) the regular school procedure. The first plan is used only in the foreman training. By this method, the group furnishes practically all the information on which all the discussion is based, the conference leader having for his function the guidance of the discussion.

The second plan was found effective in the treatment of public utility economics and transportation problems. In this plan an expert gives a talk, which is followed by a discussion of about the same length as the talk.

The regular school plan embraces the courses in English, public speaking, and the technical subjects.

"The Boston Elevated Railway, like others, is trying to raise the standards of its personnel. This, it is believed, must be done partly through education. The first aim in its program is, therefore, to foster a desire for self-improvement on the part of its employees by creating an atmosphere conducive thereto.

"The result will be not only the improvement of the individuals affected, but the management will secure information as to the fitness of individuals for promotion as adjustments of personnel are made more nearly to meet the requirements of the work to be performed."

B. The General Electric Company.—The Director of Industrial Relations of the General Electric Company states that the educational program was instituted, in general, to prepare men for specific tasks inside the company's organization. The evening courses at the Pittsfield (Mass.) plant were started in the fall of 1913 for the purpose of giving employees an opportunity to take up advanced studies in engineering, mathematics, drawing, and business administration. These courses have increased in number and importance each year, until the nucleus of a full-fledged evening technical institution has been formed. Experienced teachers, usually college trained men and women, compose the faculty. The courses are correlated to a large extent with those of the University Extension Division of the State of Massachusetts.

The courses are given immediately after work at five o'clock, and are of one or two hours' duration, leaving the remainder of the evening free for recreation. Courses in typewriting, elementary stenography, health, nursing, public speaking, elementary electricity, direct and alternating current theory, transformer design, electrical laboratory, mathematics, drawing, industrial management, properties and uses of materials, machine shop practice, vector analysis, and safety engineering comprise the curriculum.

The "foremanship training course" is particularly significant. This course is a study of the problems of supervision, given under the leadership of a factory official of broad training and wide experience, and supplemented by a related series of lectures by factory executives. Problems requiring original research and solution are introduced, and opportunity is provided for further study and assistance along lines of special interest.

In 1924 the course was limited to selected apprentices. Assistant foremen were also included. Some of the courses offered include: "Modern Production Methods," "The Man and The Job," "Handling Men," "Handling Equipment," "Principles of Management," and "Time and Effort Study."

The whole training program of the General Electric Company is representative of the better type of industrial cooperation in education. The schools of this company are progressive, modern, and, in the company's estimation, very worth while.

C. The National Metal Trades.—The National Metal Trades Association's principles and policies, as they pertain to employee training, were determined by Harold C. Smith, now president of the association. The present director of industrial education states that the necessity for employee training in the metal trades industry has been brought about by the growing scarcity of all-round skilled mechanics, also foremen and supervisors capable of dealing with the human element phase of their positions. He finds that effective employee training tends to increase the efficiency of individuals and the entire plant, and has a favorable effect upon labor turnover, the quality of the product, and the reduction of costs.

D. The Radio Corporation of America.—Mr. W. H. Beltz, the secretary of the "Student Engineer Committee" of the Radio Corporation of America, states that the Radio Corporation undertook a systematic training of young engineers in order to take care of the healthy growth the corporation is enjoying and to insure the supply of technical men trained in Radio Corporation methods and systems. He states, in addition, that the corporation's motives are identical with those of the electrical manufacturers and telephone companies which have established training courses.

E. The New York Central Railroad Company.—Mr. C. W. Cross, Supervisor of Apprentices, of the New York Central Railroad Company, indicated the basic principles of apprenticeship of the Central as follows in a letter to the author, December 3, 1928:

1. To develop from the ranks in the shortest possible time, carefully selected young men for the purpose of supplying leading workmen for future needs, with the expectation that those capable of advancement will reveal their ability and take the places in the organization for which they are qualified.

2. A competent person must be given the responsibility of the apprenticeship scheme. He must be given adequate authority, and he must have sufficient attention from the head of the department. He should conduct thorough shop training of the apprentices, and, in close connection therewith, develop a scheme of mental training. The mental training should be compulsory, and conducted during working hours at the expense of the company.

3. Apprentices should be accepted after careful examination by the apprentice instructor.

4. There should be a probationary period before apprentices are finally accepted, this period to apply to the apprentice term if the candidate is accepted. The scheme should provide candidates for apprenticeship who may be better prepared as to education and experience than is expected of the usual candidate.

5. Suitable records should be kept of the work and standing of the apprentices.

6. Certificates or diplomas should be awarded to those successfully completing the apprentice course.

7. Rewards in the form of additional education, both manual and mental, should be given apprentices of the highest standing.

8. It is of the greatest importance that those in charge of apprentices should be most carefully selected. They have the responsibility of preparing men on whom the road is to rely in the future.

9. Interest in the scheme must begin at the top, and it must be supported by the management.

10. Apprenticeship must be considered as a recruiting system, and greatest care should be taken to retain graduated apprentices in the service of the company."

The apprenticeship system was founded in order to provide skilled workmen, a number of foremen, a sufficient number of good draftsmen, and a few master mechanics and an occasional superintendent of motive power.

F. The Yale and Towne Manufacturing Company.—Mr. J. B. Chalmers, Director of the Training Schools of the Yale and Towne Manufacturing Company, states that it is rather difficult to enumerate all the factors which were the cause of the establishment of his company's apprentice school in 1908. Before that time, Yale and Towne apprentices were trained, but not in an organized school. The primary purpose in all the company's training is to have a continuous supply of trained workmen. The organized school, in addition to making this training more thorough with classroom work in trade subjects, accorded the opportunity to offer subjects closely allied to the trades, and also subjects which would raise the standard of general knowledge of the apprentices. "Therefore we might add a second purpose," says Mr. Chalmers, "that of bettering the general education of these employees and making them better citizens." Another set of deciding factors enumerated by the director include those of greater efficiency and economy obtained by segregating the apprentices under specialists.

The following statements from the "Apprentice School"

published by the Yale and Towne Manufacturing Company are significant:

"The apprentice school was established in March, 1908, to train young men in the skilled trades and thereby to maintain a constant supply of competent mechanics. The company offers many opportunities for advancement to positions of responsibility to well-trained young men of good character.

"The instruction is designed to give every apprentice definite practical skill along his chosen line of work, and at the same time develop habits of reasoning, power of initiative, and ambition. The practical work in the shops is supplemented by well-correlated group instruction in the classroom."

G. The R. H. Macy Company.—The R. H. Macy Company has spent great amounts of time and study in the fields of personnel and human relations. The sixteenth floor of its building today reveals a well-equipped educational department. The group in charge are alive to the needs of a training department, and are in possession of the latest and best thought in the modern scientific phases of the employee training movement.

The development of the germ of training at Macy's, as conceived by Miss Fike of the department, is an interesting and significant story of origin and growth. In the early days, one woman constituted the training department. Her duty was to instruct the sales people in the use of the cash check system. As the business grew and the employed staff became greater, assistants were added. The check system had become more difficult and involved as strides were made in cost accounting and improved bookkeeping. More time was required to train the salesperson in the use thereof. However, the extent of training at this time was the cash check procedure.

Later, a demand for better salespeople was felt in the store. Division heads, as well as department heads, complained of the short-comings in the sales force. High cost of sales, due to abnormal turnover and ineffective salesmanship, was noted. Turnover in itself provided no remedy. A demand that something be done with Macy help was instigated.

A training department was established to give the necessary

training in salesmanship, technique of sales, and personality. At first a few departments were selected for experimentation. The results were encouraging, and in a short time other departments clamored for sales training.

At present a complete program is in operation under the direction of a competent instruction force, which bends every effort to follow up the trained individuals.

Although no scientific study to evaluate the effect of the training program in the working of the business has yet been made, it is generally felt that the work has done great things.

All innovations, new ventures and methods set forth in the program have arisen from definite needs felt in the store.

H. The National Grange.—The National Grange today holds a high place in the estimation of the public. It is recognized as the spokesman of the American farmer. The National Grange is expressive of the best rural progress and is identified with the highest interests of the agricultural people of the nation.

The grange in 1883 definitely joined the forces demanding vocational education. In January of that year the Hon. D. Wyatt Arken of South Carolina delivered an address before the convention called by the Commissioner of Agriculture at Washington, D. C. He showed the great possibilities of educational service contained in the grange. Considering the period in which the speech was made, it is a remarkable document. In the half century since, the grange has been ever ready to recognize the values of agricultural education, and has promoted many programs of an educational nature.

I. The Westinghouse Electric and Manufacturing Company.—The Westinghouse Electric and Manufacturing Company maintains an extensive training program. In it there are divisions ranging from graduate engineering schools to simple apprenticeship training schools.

The Westinghouse Electric and Manufacturing Company has come to be one of the great national business institutions of the country. Beginning in a small way in an alley plant in Pittsburgh in 1886 with a few men actuated by the enthusiasm of George Westinghouse, it has grown until today, through the engineering ability, vision, and farsightedness of the technical graduates who have caught the spirit of the founder, its influence is felt in every field of endeavor. In the forty-five years of its existence, tremendous changes have occurred in the electric world. To the company has come the honor of placing electric lighting on a commercial basis, of developing the alternating current system, of giving the impetus to the motorization of industry through the induction motor, of making successful the steam turbine, of developing the turbine-generator, of giving electric railroading its position of prominence, and, recently, of establishing radio broadcasting. Its annual business is in excess of $175,000,000, and it manufactures over 300,000 separate products. The following is a statement from one of the brochures issued by the company, (page 5, "Engineering Graduates and Westinghouse"):

"The opportunities in electrical and mechanical engineering are limited only by the vision, ability, and courage of the young men who are now entering work."

For more than thirty-five years the Westinghouse Company has recruited the ranks of its executives and technical experts from among those who show capacity and who enter the organization directly from engineering schools.

"The company believes in the graduates of technical schools" (page 6, "Engineering Graduates and Westinghouse"). The policy of the company in dealing with the placement of technical graduates is the result of experience. It is the opinion of those in charge that the best results are obtained by coordinating in a systematic way the theoretical work of the college with practice. The "Graduate Student Course," a year of supervised practical experience, is recommended only for those young engineers who feel the need of such supplementary experience. Certain portions are recognized for credit in the Graduate School of the University of Pittsburgh toward the M.S. or Ph.D. degree.

The "Intermediate Course" for high school graduates prepares for semi-technical positions with engineering, commercial, and manufacturing departments. This course consists of two years of very practical experience supplemented with classroom instruction. The course includes shop practice, planning, scheduling, accounting, mechanical engineering, electrical engineering, sales work, and service work. The first six months are considered a probation period. A very close check is kept on the student's class and shop work to determine his fitness for the course. Those not fitted are dropped.

Courses in "Trades Training" are also fostered by the Westinghouse Company. Pattern makers, machinists, tool makers, electricians, printers, office assistants, blue print boys, and tool handlers are taught through a number of short-term courses. Each apprentice spends a minimum of four hours per week in class work. Instruction is given in mechanical drawing, heat treatment of steel, and shop problems which include English, mechanics, psychology, economics, management, shop systems, costs, and the applications of the principles of mathematics to shop work. Various positions in the company are filled from the product of these courses.

The Westinghouse Company also conducts a technical night school. There are five major departments in this branch of the employee training program.

The Engineering Department offers a four year course in fundamental engineering principles. The Junior Engineering Department gives a two year course in public school subjects and high school mathematics.

A Foreign Department maintains a two year course in written and conversational English and Americanization for foreign born men and women. The Accounting Department presents a four year course in accounting and business administration.

The Women's Department is subdivided into six divisions:

1. A three year course in commercial subjects.
2. A two year course in sewing.
3. A one year course in calculating machine operation.

4. A one year course in preparatory training.
5. Special courses in dictaphone operation and fast dictation.
6. Special courses in handicrafts and cookery.

The bulletin of the "Westinghouse Technical Night School," page 7, states:

"To encourage young men and women to devote a portion of their spare time to educational training and development with the aim of equipping them to take their proper place and assume greater responsibilities in society and to provide that which best develops strong character are the highest aims of the Westinghouse Technical Night School.

"Since its organization in 1902, the school has passed through many stages of development. At its inception it was purely a corporation school. It has, however, gradually broadened its scope and extended its services to include added courses of training in order to bring increased benefits to both the community and the industry. Today the school offers varied courses of training to men and women.

"To young men and young women the school offers an opportunity to secure advanced education which will fit them for more responsible positions in life. It offers to those who must work during the day a plan whereby their leisure hours can be utilized to the greatest possible advantage. It offers to them a means of making themselves better citizens, better employees, and more successful men and women.

"To the local industries the school owes its early existence, and even now depends upon them for substantial support. The school offers to the industries a place of training for their employees, and produces a trained worker better able to meet the requirements of industry.

"The local school boards, believing generally in the cause of education and in the type of training offered by the school, contribute in a large measure to the financial program of the school. To the local communities the school offers an opportunity to raise the standard of their citizens and to help in training men and women for a higher type of citizenship."

J. The National City Bank.—A press clipping from the New York Herald Tribune of March 12, 1929, indicates the attitude of one of the biggest banks in the country toward education.

"An educational trust fund, designed to enable employees of the National City Bank and its affiliated corporations to pursue higher practical and cultural education than their own means would make possible, has just been formed by voluntary subscription by a group of the bank's officers of a sum, immediately available for work, equal to the income on more than $2,000,000.

"The National City Foundation is in no sense designed to take over the educational work that has already been developed within the National City organization, but rather to supplement that program."

Mr. Charles Mitchell, president of the bank, said in commenting on the project:

"The thing that has always impressed me about the educational training offered young employees by corporations and banks is that its purpose is essentially to benefit the institution rather than the individual specifically. Necessarily, this is because the distance to which a corporation properly may go in the matter of employing its funds for educational activities must always remain an open question.

"The principle upon which the foundation is based—the creation of educational facilities through bequests and the establishment of trust funds—is not new, but its application is new. Large amounts of money are bequeathed every year to educational institutions, but the benefits of these bequests go solely to those with time and funds to attend the institutions receiving the money. The National City foundations, in contrast, will offer opportunities heretofore not available, to the bank's employees, and this offer is not conditioned upon their having time to spare to devote to study or funds to pay for the education they receive. The bank will see that they have the opportunity, and the necessary funds will be provided by the foundations."

Summary.—These are but scattered pictures of the interest organized business has taken in employee training. The results have been no mean contribution to the evolution of organized trade training. Along with the development in industry and in business has come the development of vocational education. The one is inextricably tied up with the other. Their origin has been in the same place, in the need for better goods, more efficiently produced; in the need for more efficient workers, who are at the same time better members of society.

Part Five

VOCATIONAL EDUCATION IN CANADA

Part Five tells the story of vocational education in Canada. Canada has had the advantage of observing the principles and practices of vocational education being applied in England and the United States.

In the construction of this part of this work the bulletins issued by the Canadian government on "Vocational Education" have been perused. Especially significant is Bulletin No. 28 issued by the Department of Labor of Canada on "Vocational Education."

VOCATIONAL EDUCATION IN CANADA

Canada Has Had the Opportunity of Witnessing the Industrial Development of England and the United States. —It is interesting to study the development of vocational education in Canada. For several reasons such a study is significant. The Industrial Revolution came to Canada long after its effects were felt in the United States. Canada has had at its disposal the history of vocational education in Europe and America. Canada has had the opportunity to benefit by the mistakes and shortcomings of educational projects in the more advanced industrial and agricultural countries.

Early Discussions of Vocational Education in Canada. —It is difficult to know just when vocational education was first discussed as a desirable type of training in Canada. Possibly the earliest settlers recognized the need of a special training which would enable them to cope with conditions in the new country. About the time of Confederation (1867) the Hon. Mr. Carling, Commissioner of Agriculture for Ontario, stated in a report:

"Notwithstanding the great advancement we have made within a period comparatively short, I have a growing conviction that something more is required to give our education a more decidedly practical character, especially in reference to the agricultural and mechanical classes of the community, which compose the great bulk of our population, and constitute the principal means of our wealth and prosperity. What now appears to be more especially needed in carrying forward this great work is, in addition to the ordinary instruction in the common schools, the introduction of elementary instruction in what may be termed the foundation principles of agriculture and mechanical science." (Quoted on page 1, Bulletin #28, Dept. of Labor, Canada, entitled, Vocational Education.)

Egerton Ryerson, Chief Superintendent of Education for Ontario, in his annual report for 1871, expresses similar ideas:

"Technical education is instruction in the peculiar knowledge or special skill required in any business or occupation, the training which will render the talents of the citizens most useful to the state in that particular craft or profession in which he or she is engaged, whether mechanic, farmer, engineer, teacher, merchant, architect, minister, doctor, or lawyer. As the education of the common school fits the youth for the performance of his duties as a citizen, so the technical school prepares him for the special duties of his trade or profession. Divinity, law and medical schools for special or technical instruction have long been in successful operation." (Quoted from same Bulletin).

Professor Walter Smith of the Conservatory School of Fine Arts, Boston, Mass., gave an address before the Council of Arts and Manufacturers of the Province of Quebec at Montreal. In the course of this address, Professor Smith said:

"Private enterprise and munificence are at work for the benefit of the public, and Mason and Holloway are just now starting on their great careers of usefulness, establishments for the promotion of technical education which dwarf by their extent and magnificence even national action in the same direction.

"It is time, therefore, for every country to decide whether it can afford to stand still and watch this movement only and do nothing for itself. When the whole world is moving, the stagnant country is rapidly drifting to leeward, and will soon find itself out of the race of progress. I beg to draw the attention of the publicists of the country to the statement which I have many times made on the other side of the frontier, namely, that it is not for the benefit or the happiness of a people to rely wholly on agriculture as a means of support. There is no profit and no honor in being the hewers of wood and the drawers of water for the skilled nations; no prospect of improvement in continuing to provide for them the raw material of the arts at a low price, and purchasing from them the manufactured goods at a high price. We employ six men to raise cattle, corn, coal, oil, lumber for the skilled people, and they send us back some of our own products turned into wealth to pay for our raw material, in the proportion of the labor of one man who works with skill paying for the labor of six men working without skill.

"We cannot shut ourselves out from the rest of the world in these days. They will compete with us whether we like it or not. The world is a very small place nowadays, and a manufacturer in Paris and London and Berlin is this moment competing with one of our own manufacturers in the next street, and will beat him because he is more skilled, has better workmen, has a more steady demand from a cultivated public for his goods, and can therefore afford to put more work, skill, and beauty into them, than we can, or even know how to do. We are assembled in Mechanics' Hall, and we realize that mechanics' institutes, young men's institutes, evening drawing schools, and some technical classes have long existed in the several provinces of the Dominion, and have doubtless here as elsewhere done great good. But permit me, as one who has watched this great question in another country similarly circumstanced to this, to say to you with all frankness and good intent, that these private or semi-public organizations will not provide a national remedy for a national deficiency. The question is too great, the matter at stake too important to trust it to the efforts, usually feeble and often intermittent, of the few." (Quoted from page 2, Bulletin #28.)

These few quotations are sufficient to show that the need for vocational training was recognized by leaders in industry and education in Canada long before work in vocational education was introduced into the secondary schools.

Early Attempts at Vocational Education in Canada.—It is difficult to trace the beginnings of vocational education in each province, but it may be said that, with rare exceptions, vocational work in the day schools was not organized until early in the present century. Evening classes were started in the 'eighties and commercial classes in the day schools were in operation at several places before the close of the last century.

In Nova Scotia the first evening schools were especially designed for miners. In reviewing the history of the mining schools of Nova Scotia the Canadian Mining Journal for July 21, 1922, states:

"It was in the year 1866 that the matter of forming mining schools in the collier districts of Nova Scotia was first discussed. The Hon. Robert Drummond, who was then Grand Secretary of the Provincial Workmen's Association, was instructed to interview the Nova Scotia

Government on the subject of mining education. Now, as the coal areas of the province are retained by the Government and leased to operators for definite periods, and as the largest part of the provincial revenue is derived from this source, it was not difficult for a man of Mr. Drummond's ability to persuade the Government that night schools were a necessity. He was able to point out the advantages of having the mines under the control of miners well educated and trained for their positions. Safety, efficiency, and sympathetic management were sure to come as a direct result of an educated class of native workers. And the long period of unbroken harmony which existed at most of the Nova Scotian collieries amply justified Mr. Drummond in his argument, and the Government in acceding to his request in opening the schools.

"In 1888 the first mining schools were opened. To secure suitable teachers was the first difficulty to be overcome. The schools were placed under the mines department. For a time in some centers instruction could only be given to enable students to acquire second and third grade certificates. Those desiring first class certificates had usually to plod alone or with such help as they could get from day school teachers and others outside the mining profession. As the first local instructors were men who had to work hard all the day, their teaching efficiency was not of the highest, and the government, after reviewing what had been accomplished since the schools were first opened, decided to establish a technical college in Halifax. This college was opened in 1907 and took over the mining schools."

Modern Periods of Canadian Trade Training.—Since 1900, the history of secondary vocational education in Canada has been eventful. The story divides itself into three ten-year periods. The first includes the years 1900 to 1909, preceding the appointment of the Royal Commission on Industrial Training and Technical Education; the second is from 1910 to 1919, preceding the enactment of the Technical Education Act; and the last is the ten-year period of the act's duration, which terminated March 31, 1929.

Canadian Vocational Education From 1900–1909.—During the 1900–1909 period efforts were made to introduce manual training and domestic science into Canadian schools. Due to the personal interest of Sir William MacDonald, especially qualified manual training instructors were brought to Canada from Great Britain and placed in representative schools throughout the Dominion. Through the efforts of Mrs. Lillian Massey-Treble of Toronto, and Mrs. Hoodless of Hamilton, domestic science was introduced into Ontario schools. Such efforts on the part of private individuals encouraged the provincial departments of education and local school boards to establish practical work in the school systems, and started the movement which has resulted in the present development of vocational education in the Dominion. Manual training and domestic science, which were pioneer subjects in the development of secondary vocational education, are now usually found as established parts of the academic branch of school work and are, in consequence, not included in vocational education programs.

Nova Scotia was the first province to organize a program of vocational education of less than college grade. In 1907 the legislature passed the Technical Education Act which provided for centralized control of all technical or vocational education in the province. A provincial director was appointed, and provision was made for the inspection and supervision of evening schools throughout the province. As has been stated, Nova Scotia had conducted evening classes for coal miners since 1888. Under the new act, the scope of these classes was enlarged and provision made for other types of work.

The first day vocational school in Canada to be operated in a separate building was opened in Toronto in 1901, and the first building erected specially for secondary vocational education was erected in Hamilton in 1909.

The appointment of the Royal Commission on Industrial Training and Technical Education in 1910 resulted from or perhaps merely coincided with, an active interest in vocational education thoughout the whole Dominion, and marked the beginning of the present development.

The commissioners, after carefully studying the existing conditions and requirements in Canada and inspecting the work being done in the United States, Great Britain, and parts of Europe, recommended a development policy and complete system of vocational education for Canada. This system, although retaining provincial control of education, provided for federal, provincial, municipal, and private financial contributions, and close cooperation between schools and industry.

The types of work suggested for urban communities, as quoted in Bulletin #28, Vocational Education, page 4, were as follows:

"For those who are to continue at school, intermediate industrial classes; coordinated technical classes; technical high schools; apprentices' schools; industrial and technical institutes; and home economics and fine arts colleges.

"For those who have gone to work there should be continuation classes, coordinated technical classes; middle technical classes, apprentices' classes in work shops, industrial and technical institutes, and correspondence study courses."

Similar provisions were suggested for rural communities. The commissioners also recommended that the sum of $3,000,000 be provided annually for a period of ten years by the Parliament of Canada and paid into a Dominion Development Fund to be used for teachers' salaries, for the establishment and maintenance of classes, schools, etc.; to provide suitable equipment; to make provision for scholarships; to pay salaries of experts with experience in industrial training and technical education, and to promote scientific industrial and housekeeping research and the diffusion of knowledge resulting therefrom.

While the Royal Commission for Canada was making its investigation, the province of Manitoba appointed a commission to make a study of vocational education in Canada and the United States. The results of the study were epitomized in the following recommendations, quoted from page 4, Vocational Education, Bulletin #28, Department of Labor, Canada.

"1. That it is desirable that such measure of vocational training as is found possible should be provided for the people of our province.

"2. That the foundation for such training should be laid in the elementary school in suitable courses of hand and eye training, leading up to regularly organized industrial work in the higher grades of the elementary and through the secondary school.

"3. That vocational and general education should go hand in hand, each in turn contributing to the effectiveness of the other, and each recognizing their interdependence.

"4. That at the present stage of our development this can be done more economically and effectively by the modification of existing agencies and the enlargement of their scope than by establishment of special institutions.

"5. That a certain number of the members of the Advisory Board should be men engaged in the industries and selected on account of their acquaintance with and interest in the aims and ideals of vocational training; and that school boards be authorized to appoint advisory committees outside of their own members to assist them in the organization and development of the work of vocational education.

"6. That school boards be authorized and encouraged to provide such courses in vocational education as will suit the needs of the respective communities.

"7. That such provision should include carefully organized evening classes, in which opportunity would be given to men and women engaged in occupations during the day to improve their general and technical education.

"8. That the Department of Education should appoint an officer familiar with the aims and methods of vocational education, whose duty it would be to advise with and assist school boards in the organization of such work.

"9. That grants be made by the Provincial Government, as is now done in the case of manual training and household science, to assist in meeting the cost of equipment and maintenance of approved lines of vocational training.

"10. That in any scheme of education looking to increase vocational efficiency, provision must be made for systematic physical education.

"11. That provision be made for the preparation and training of teachers to meet the requirements of the new activities of the schools.

"12. That when there shall be a sufficiently large number of students requiring higher training than is herein provided for, a technical college shall be established to provide such training."

Dr. Seath's Visit to the United States.—During 1910 Dr. Seath, Superintendent of Education for Ontario, was sent to the

United States and to Europe to study vocational education and to suggest methods of developing work in that province. After an extensive survey Dr. Seath offered a report containing a series of recommendations touching upon all phases of vocational education. The report indicates a fundamental belief in the principle that a general education is an essential preparation for all vocations and is made more effective through a closer connection between the schools and the activities of life. The extensive report did much to stimulate thinking and action in regard to vocational education.

Canadian Legislation on Vocational Education.—The Royal Commission appointed by the Dominion Government made its report in 1913, and in the same year Parliament passed the Agricultural Instruction Act, under the provisions of which the sum of $10,000,000 was set aside to promote and assist agriculture and agricultural instruction during a ten-year period. This money was paid to the departments of agriculture and used to develop new branches of agricultural work which would directly and indirectly assist the farmers in their work and home life. Part of the money was used for school purposes in promoting agricultural education, but most of it was spent in connection with demonstration work, clubs, and experimental stations. The act expired in 1923 and was renewed for one year to give the provinces a chance so to adjust the work that they could assume the full financial burden. As this was considered to be impossible by some of the provinces, a considerable part of the work was discontinued and much of it was never revived.

The recommendations of the Royal Commission on Industrial Training and Technical Education were not acted upon until 1919, due to the advent of the World War. But in 1919 the Technical Education Act was passed, which provided $10,000,000 to be set aside for use in the promotion and development of technical or vocational education for persons entering or employed in industrial pursuits.

According to the provisions of this act the Canadian Government assists in the promotion and development of vocational education of less than college grade by sharing equally with each provincial government approved expenditures on all branches of vocational education. In order that the smaller provinces may earn a fair share of the grants, the sum of $10,000 is allotted to each province before dividing the balance of the annual appropriation in proportion to population.

Canadian Vocational Education Since 1919.—Since 1919 the development of vocational education has been rapid, and in no two provinces has the evolution been the same. The statistics and written reports of the provinces give only a surface picture of the changes, and do not indicate the peculiarities and local changes which are characteristic of any system of vocational education.

A few figures might aid in picturing Canada's vocational system: In the year 1919–1920, there was a total of 1,810 teachers engaged in teaching vocational subjects (this includes day, evening, and correspondence school instructors) and 60,546 pupils. During the year 1926–27 there were 3,660 teachers and 96,682 pupils (Table I, Bulletin #28, Dept. of Labor, Canada, Vocational Education).

Quebec's Program.—Quebec's system of education is unlike that of any other province. Provision is made for instruction in both French and English, and a large majority of the pupils are French speaking. The Council of Arts and Manufacturers, incorporated in 1872, initiated the first system of evening schools in the province of Quebec. In 1906 the Council of the Montreal Board of Trade took an active interest in the work being done, and urged the government to increase the grant to the Council of Arts and Manufacturers.

Montreal's Program.—The Montreal Technical Institute had its origin in the Canadian Manufacturers' Association, as a result of repeated complaints to the executives of the association from manufacturers respecting the unsatisfactory condition of technical education in the city. A committee took up the subject with the Mechanics' Institute and some influential citizens, and a charter was obtained for the institute. In 1908 the Protestant Board of School Commissioners agreed to cooperate with the

Institute to the extent of granting the free use of the Commercial and Technical High School building for evening classes, and of assuming the management of the course of instruction outlined by the institute.

Two provincial schools of art in Quebec and Montreal were opened in 1922 and 1923 respectively. Over 1,500 pupils are enrolled in day and evening classes in these schools, which teach various branches of pure and applied art. Commercial education, which receives federal grants, is confined to certain branches of the work carried on by the School of Higher Commercial Studies, in Montreal. In addition to degree work in commerce, this school provides an extensive evening school program and correspondence courses for office workers and others engaged in commercial work. A pulp and paper school was established in Three Rivers in 1923. It is operated by the Department of Lands and Forests as a branch of the technical school, and provides full-time, part-time, and evening courses for workers in the pulp and paper industry. The part-time and evening pupils are engaged in the local plants. A provincial school for forest rangers has been in operation in Berthierville since 1924. It is also operated by the Department of Lands and Forests.

A printing school for apprentices and journey men was established in connection with the Montreal Technical School in 1925. This school is supported by the provincial government, but it is operated by an advisory council composed of representatives of the printing employees, the employers, and the provincial government. A similar school for the building trades has been under consideration and has since been opened.

The Shawinigan Technical School was built and equipped by a local industry, and, although it receives federal and provincial grants, it is supported mainly by the industries which it serves. The schools at Grand Mere, La Tuque, and Sherbrook are taught by religious orders, but are open to any boy in these municipalities. The Council of Arts and Manufacturers conducts evening classes in fifteen municipalities. All the abovementioned institutions are recognized as a part of the provincial vocational education system.

Vocational Education in Ontario.—In Ontario the story is similar. The need for training in the higher technical fields was recognized as early as 1870, and since then vocational education has crept down into the secondary and elementary schools.

Summary.—From 1870 to 1900 there was a marked movement for the establishment and development of many agencies for vocational education of a professional type. Opportunities for specialization were provided for those entering the skilled occupational fields before they were provided for those looking forward to the semi-skilled and non-skilled fields.

A second period covers the years 1900–1912. This was a period of discussion about the value of hand training, and of the need for providing special training for workers in the less technical fields of employment as adequate for their needs as had already been provided for the professional workers. Manual training and domestic science were introduced, and spread rapidly through the school system.

Dr. Seath's report was followed by legislation in Ontario. In 1911 the Industrial Education Act was passed. In 1913, 1915, and 1921 further acts were passed, carrying out Dr. Seath's recommendations.

The third period of Ontario's vocational education development covers the years 1912–1919. This was a period of little progress, as the World War seriously retarded the advance of education.

After the war, and to the present time, there has been a rapid extension of vocational education. Two factors played an important part in this extension. First, with the close of the war local school authorities were again in a position to deal adequately with educational problems, and in particular with the matter of deferred building programs. Second, the grants from the Dominion Government, made possible through the Technical Education Act of 1919, and increased by the provincial legislature, enabled the Department of Education to pay grants to local school authorities on capital expenditures for buildings.

"Ontario's program of vocational education is the most diversified and highly organized in Canada. Cultural and academic subjects are given first place in vocational schools and no attempt is made to produce skilled workers in any branch of industry.

"A broad foundational training is aimed at, rather than specialized training in any one branch of industrial or commercial life. The schools endeavor to maintain close contact with industry through evening classes, part-time and continuation classes, employment service for graduates, and occupational information. Apprenticeship is encouraged, and in December, 1926, the Construction Apprenticeship Council for the building trades was established in Toronto" (page 16, Vocational Education, Canada Department of Labor).

Part Six

VOCATIONAL EDUCATION AND THE FUTURE

Part Six is concerned with vocational education and the future. The purpose of the entire work has been to dig out the roots of vocational education of the past so that light might be thrown on current practices and future trends. The age-old adage that one cannot understand the future without a knowledge of the past is neither untrue nor outworn. The purpose of this section is not to predict what will occur in the future in vocational education, but rather to indicate what seems to be a logical development based on past experiences.

VOCATIONAL EDUCATION AND THE FUTURE

Change, the Characteristic of the Modern Age.—One of the salient characteristics of the modern age is change, constant and rapid change. In fact, the modern temper accepts and understands change with a freedom and ease that is vastly different from the smugness and conservatism of recent eras.

The sweeping and radical changes in industry since the Industrial Revolution have been noted. Industry, however, has not been the sole sphere of change. Business methods, practices, and attitudes also have changed. Social life, manners, and customs of today are extremely different from the social life, manners, and customs of the Victorian era of sacred memory. The arts, both fine and commercial, have also felt the influence of modern ideas and conceptions. This spirit of change and progress has even been reflected in the many and varied types of social legislation fostered by the sundry governing bodies of the land.

Predictability, a Business Necessity.—Many studies of the past have been undertaken solely for the purpose of furnishing a basis for prognostication. Business cycles have been figured out with uncanny accuracy. Trends in fashions have for years been predicted by the seers of the fashion world. Change is capricious, however, and often the predictions are found to be untrue and valueless. How often are the onward and upward prognostications brought to naught by the downward slumps of reality!

Industrial Education and the Future.—Industrial education of the future will be a response to the needs of society as it has been in the past. Its conduct and current will find its source in industry. Industrial education of the future will be an answer to the crying needs of industry. As industry changes, so education for industry must change. As new industrial developments are found to serve society more efficiently, so educational practice must adjust itself to find place for these modifications within its structure.

Three Trends of Modern Industrialism.—In spite of the futility of attempting to paint the picture of industry of the future, a few trends which appear to be universal and constant may be mentioned. In the first place, there appears to be a steady increase in the specialization of the processes of manufacture. This specialization had its root in the Industrial Revolution, and since that time has increased and seems likely to continue its increase. This event, of course, will be accompanied by a further nicety in the division of labor. The recent practices and recommendations espoused by the leading engineering societies and industrial groups point to an era of intense specialization.

The second marked trend closely associated with the first, is toward standardization of products. Perhaps to the automobile industry, more than any other one branch of industry, is due this modern practice of standard parts. All industry has followed, and the result has been greater efficiency in the replacement and repair of machine parts. This standardization appears to be a fundamental part of future industrial activity.

A third trend in industry is showing itself more plainly day by day. This trend is the concentration and localization of production centers. Mergers, unions, and working cooperations have occurred in many of the leading industries of the country. General Motors, The New York Central Railroad Company, and the huge bank mergers are only a few of the gigantic enterprises of a cooperative nature. This tendency has resulted in geographic localization of industrial plants. There is a "Fashion Park" in Rochester, N. Y., an automobile center in Detroit, a film colony in Hollywood, and so on throughout the country.

The first two trends are bound to affect the general character of vocational education. The third will undoubtedly affect the choice of types of training that any given community will choose to offer.

Fundamental Technical and Trade Training, a Necessity for Employment in Industry.—Industry, probably for some time to come, will demand a more general technical and mechanical training system of the public schools. This will mean a reduction in divisions which seek to give highly specialized applications and skills. With the intense specialization of machinery and process, industry will attempt to induct its recruits by means of vestibule schools and shop courses, provided these recruits have the basic foundation of general industrial training. This seems, then, to point to an extension of the industrial arts program in the junior and senior high schools. Also, it would indicate an expansion in the low-skilled and semi-skilled departments of labor. This does not indicate a disappearance of the training for the skilled workers; on the contrary this will have to be continued with even more minute specialization, most of which will be conducted by and in the shop.

Higher Technical Training and the Future.—Institutions of higher learning will be forced to offer broad training supplemented by cooperative work in industrial and business plants. The engineering student of the future will, no doubt, be a cooperative student. He will probably spend a portion of the year in his classroom learning the general bases of industrial practices and the theories underlying his profession. The rest of the year he will find himself in an industrial shop or business office carrying out the principles he has learned at school, and working out the practical applications of his theories. College laboratories then will be actual shops and, as such, progressively equipped with the latest and most efficient devices.

Localization of Industries, a Trend of Modern Business.—The third trend, that of geographic localization, will unquestionably affect the local school district's choice of vocational program. It will then be necessary for local boards to transport students to other districts when they elect a trade or course which their district deems too limited in appeal for local adoption. Great schools will probably grow up about these centralized and specialized industrial units. To them, it may be expected, hosts of young people will flock, provided they are assisted by their local school boards. Suburban communities and residential areas then may take advantage of schools offering specialized curricula in adjacent districts and contribute to their upkeep. This in turn, seems to indicate a growing tendency to make vocational education a district, county, state, or possibly even national function in order to overcome the lacks and inefficiencies of localized puny efforts. The agitation for a Westchester County system of vocational education in New York State is an example worth considering.

Improvement in the Administration of Vocational Education Must Come.—As for developments in vocational schools, we may expect further strides in the perfection of administration. Research, carried on by the Federal Board for Vocational Education as well as by students in schools of education, will, no doubt, reveal ways and means of improving and enriching the administration of the vocational schools. Already the realization is gaining ground that a vocational school does not lend itself to the same administration as the academic high school. Directors of vocational education, it seems, will be made associate superintendents, so that they may have easy access to the legally constituted boards of education.

Improved Teacher Training, a Characteristic of Modern Education.—With the establishment of vocational education departments in the various schools of education and the further provisions for improved teacher training, we may expect better qualified teachers of vocational subjects. The influence of the teacher upon his students and the school system is great. In the past, as has been shown, many great movements in vocational education received their impetus from great men. Capable teachers who can make use of the latest in scientific methods and practices will revolutionize the procedure in vocational schools.

The Vocational Teacher Achieves a Place in Society.—Within a very short time there will have grown up a generation of teachers who have been trained professionally and technically to be teachers of vocational subjects. This will end for ever, it is hoped, the reign of the vocational teacher whose background and experience have been purely academic and theoretical.

It is gratifying to notice also that the old feeling of hostility between academic and vocational teacher is dying down. Too much light has been thrown on the subject of "culture" in recent times to permit a continuance of the smug feeling of superiority.

The Vocational Student of the Future.—The vocational students of the future will undoubtedly be a better-selected group. Applied psychology has contributed immensely in the way of standard tests and measurements. Vocational guidance has made such strides that fewer and fewer of the vocational school population will find themselves misfits, and, consequently, unhappy members of society.

Summary.—Where will the impetus for vocational education come from in the future? It is reasonable to expect that the forces, whose roots in the past have made possible vocational education, will continue to direct the course of this practical education.

There is already evidence of mutual cooperation among the forces of industry, education, and labor to promote more efficient training for young workers. Social institutions, private and public, have their contributions to make. These contributions, however small, may be joined to those of industry, education, and labor—advances all of which can be cemented by legislative enactments and appropriations.

In the end it is hoped that vocational education, the possibilities of which are so tremendous, may make its contribution to the future happiness and progress of the race.

Although the emphasis of this part of the work may appear to rest wholly on the subject of the full-time programs for vocational training, it is to be realized that the present trend in industrial education is toward training for employed youths and adults. Modern developments in short unit-courses and in the continuation school field are little short of startling. Vocational education is truly assuming its rôle as a type of efficient "pusher education." The large numbers of institutions which invite students to enter for short, definite courses bear witness to this recent trend.

The roots of this type of training are not difficult to discover. They are buried in the complex social structure of the modern age. The social and economic order demand that all save a favored few engage in gainful employment at an early age. The result has been the great movement for part-time education. Learning has gone hand in hand with earning. At the same time workers in service have noted the increased wage paid in more advanced positions. Continued education has been the result here. The ever-changing machines and processes of present day industry create new jobs and methods. Workers have found that they must educate themselves continually in order that they may keep abreast of the times. Short unit courses and opportunity groups have been presented to meet this demand.

Part Seven

MILESTONES IN THE ADVANCE OF VOCATIONAL
EDUCATION

MILESTONES IN THE ADVANCE OF VOCATIONAL EDUCATION

Introduction.—The growth of vocational education has been slow, as almost all healthy developments of an evolutionary character have been. The origins of vocational education can be traced back to remote antiquity. The father of vocational education was, without doubt, the first savage who taught his fellow some simple thing connected with the struggle for existence—the making of an arrow, or the catching of a fish, for example.

It is the purpose of the following table to indicate a few of the striking milestones along the path of progress in vocational education. No attempt has been made to cull from history all the causes that have gone to make up vocational education. Such a problem would give all history as its field, for developments in one field can rarely be separated from developments in adjacent fields. The organic development is universal, catholic; and reactions in one part are felt throughout the whole.

With these thoughts in mind, the table has been constructed. Three parallel columns give developments in political-legislative, economic-social, and educational-philosophical history.

Year	Political-Legislative	Economic-Social	Educational-Philosophical
Pre-Christian Era;, 2500 B.C.	1. Egypt Feudal Age. 2. The Empire (1580–1150).	Irrigation systems perfected on the Nile.	Egyptians had vocational education for entrance into the sacerdotal class (priests, teachers, administrators), the military class, and the industrial class. The apprentice system was used.
Up to 500 B.C.	The Persian Empire.		Persians had systematic education for those who were to be Magi, soldiers, and artisans.
			Genesis (4, 22) mentions Tubal-Cain, an instructor of every artificer in brass and iron. The first teacher in a vocational school.
1500 B.C.–930 B.C.	The Hebrew Kingdom.	Tyre and Sidon were important trading cities of the Phoenicians.	Hebrew Talmud demands the teaching of a trade. Hebrews established elementary schools featuring moral and religious education.
1500 B.C.–70 A.D.	The age of the Greek World.		Spartan education for the development of the individual as a soldier of the state.
776 B.C.–338 B.C.	The Golden Age of Athens (Pericles) 338 B.C. Philip of Macedon gained leadership of Greece.		Athens had great leaders in education. Socrates (469–399 B.C.) developed theory of concepts and the question method.
334 B.C.–46 B.C.	The Age of Alexander and the union of Greece with the Orient.	Oriental products filled European markets. International commerce and communication developed due to products of the East finding great favor in the West.	Reasoned from ignorance to truth. Plato (427–347 B.C.) developed idea that men and women should do the work for which they are best fitted. He set as the goal, virtue in the individual, justice in the state.
			Aristotle (386–322 B.C.) was Alexander's teacher. The function of the state, he said, is the art of so directing society as to make for the greatest good of mankind.
50 B.C.–500 A.D.	The Age of Rome. Julius Caesar had conquered the civilized world by 44 B.C. and reorganized the government.	Caesar's military successes were reflected in Roman trade supremacy, and commerce flourished by sea and over the famous Roman roads.	Roman education was practical. It featured military, legal, religious, and technical training.
	In 33 A.D. Jesus Christ was executed by the Roman governor of Palestine.		Christianity gave a philosophy of human brotherhood and welfare of all through welfare of individuals.

Year	Political-Legislative	Economic-Social	Educational-Philosophical
330 A.D.	Constantine moved his capital from Rome to the new city of Constantinople.	Eastern influences dominated the industries of Rome.	Education became "cultural."
500–800	European Dark Ages and loss of past educational gains and commercial advances.		
650–1200	Period of Saracenic Culture.		Saracens made great strides in science and mathematics.
900–1400	The Feudal Age.		Chivalry—the school of the knight. Cathedral schools.
600–1200			The monasteries conducted schools for (1) entrance into Holy Orders, (2) for the arts (3) for agriculture.
800	Charlemagne, King of the Franks.		Alcuin of York together with Charlemagne led an educational revival.
875	Alfred, King of England.		Alfred established the Palace School at his court.
1066	Norman Conquest of England.	Continental influence felt in England's economic life.	William brought the English church in closer relationship with the Roman church.
1080			Salerno, a health resort in Italy, established a university specializing in medicine.
1100–1300	Age of the Crusades.	Revival of interest in Oriental manufacture and articles of the east.	The learning of the Orient was a revelation to those who came in contact with it.
1158			University of Bologna founded to teach civil and canon law.
1180			University of Paris, adjacent to the Cathedral of Notre Dame, founded to teach theology; Abelard taught there.
1208			Francis of Assisi founded the Franciscans.
1216			Dominic founded the Dominicans, who established themselves at the university centers and gave advanced training. The preachers at the courts were Dominicans, as were the preachers of the Crusades. These include Roger Bacon, Scotus, Vincent de Beauvais and Thomas Aquinas.

Year	Political-Legislative	Economic-Social	Educational-Philosophical
1257			Thomas Aquinas produced the "Summa Theologica," perhaps the best exposition of the theology of the Church of Rome.
1300–1500			The guild schools of the late Middle Ages trained vocationally for mastership in the practical arts. They established elementary schools for the children of the craftsmen.
			Chantry schools were established with the funds left by the wealthy for masses, the priests to teach children during their spare time.
1215	Magna Charta signed in England.		
1264	Simon de Monfort established Parliament in England.		
1304–1374			Petrarch attacked Aristotle, existing universities, and scholastic learning. He collected and copied the works of Cicero and secured the establishment of humanistic studies at the University of Padua. He wrote works in Latin to make the great men of antiquity real to his generation.
1349		Black Death swept England and depleted the labor supply.	
1381		Peasant's Revolt in England.	
1350–1400		England turned to the wool industry.	
1350–1600	The Renaissance spread through Europe from the period of the Moslem conquest of Constantinople until the period of the Protestant Revolt. Its effects were felt in the political, economic, social, educational, and philosophic fields.		
1400			John Huss executed for his religious beliefs.
1450			Establishment of first chair of Greek north of the Alps (University of Paris).
			Establishment of the Vatican Library in Rome.

THE ADVANCE OF VOCATIONAL EDUCATION

Year	Political-Legislative	Economic-Social	Educational-Philosophical
1467–1536			Erasmus, northern humanist, wrote the "Liberal Education of Children"; and "The Right Method of Study."
1485	Henry VII established the Tudor Line in England.	Henry replaced the local control of industry with national control.	The Tudors aided university education by establishing several colleges. (Trinity, etc.).
1465–1536			Erasmus wrote "In Praise of Folly" in which he ridiculed the practices and beliefs which Luther attacked. He advocated the New Testament as a book for all to read in the vernacular.
1492		Columbus discovered America.	
1509–1547	Henry VIII, King of England.		Henry VIII destroyed the monasteries.
1517			Luther published the 95 theses.
1534	Act of Supremacy declared Henry VIII head of the church in England.		
1520–1700	The Age of the Protestant Revolt.		
1540–1700	The Age of the Catholic Reaction.		
1540			Ignatius Loyola founded the Jesuits to carry on educational work.
1545–1563	The Council of Trent of the Roman church met to consider the Protestant defection. The Council ordered the reopening of the parish schools and the systematic training of the clergy.		
1553			Rabelais died. During his life he attacked the traditional beliefs through his satiric writings.
1558–1603	Elizabeth ruled over England.	Tudor autocracy placed England at the forefront of the nations of the world.	
1561			Francis Bacon born.
1598	Edict of Nantes, Calvinists in France allowed to hold church services.		

Year	Political-Legislative	Economic-Social	Educational-Philosophical
1601		Poor laws, apprenticing paupers, England.	
1605			Bacon's philosophy of realism formulated.
1607	Jamestown, Virginia founded.		
1616			Shakespeare died.
1619		James River furnaces for manufacture of iron established.	
1620	Pilgrims landed. Slavery introduced to America.		Bacon's "Novum Organum" written.
1626			Bacon died.
1630			Franciscans established industrial mission schools in New Mexico.
1630	Boston founded.		
1630 1633 1636		Massachusetts general court passed acts to regulate trades and limit wages.	
1643		Skilled weavers from Yorkshire, England settled at Rowley, Mass., and set up looms and fulling mill.	
1643		John Winthrop established a blast furnace at Lynn.	
1647			School ordered for every town of 50 households in Massachusetts.
1648	End of Thirty Years War.		
1649	Charles I beheaded in England and commonwealth established.		
1656		Massachusetts required every family to make at least 3 lb. of cotton or woolen yarn weekly for thirty weeks in the year.	
1662		Virginia legislature passed an act requiring a tannery in every county.	
1665	Great Plague in London.		
1667			Spinoza died. Moxon began to publish "Mechanick Exercises," England.

THE ADVANCE OF VOCATIONAL EDUCATION

Year	Political-Legislative	Economic-Social	Educational-Philosophical
1681	Charles the Second of England granted the Commonwealth of Pennsylvania to William Penn. In the grant, or "Great Law," Penn had inserted the provision that all youth over twelve years of age should be taught some useful trade or skill, to the end that none might be idle, that the poor might work to live, and the rich, if they became poor, might not want.		
1684			La Salle founded Christian Brothers to provide Christian education for the children of artisans and of the poor.
1685	Revocation of the Edict of Nantes.	Emigration of over 50,000 families from France to England and Holland. These included, for the most part, artisans and men of letters due to revocation of the Edict. Spitalfields colony, near London, made by the French weavers.	Budd offered a plan for public schools in Pennsylvania and New Jersey.
1690			John Locke published "Essay Concerning Human Understanding."
1692			William and Mary college founded.
1693			Locke published "Some Thoughts Concerning Education."
1694			Francke—Institute for Orphans, Germany, founded.
1694			Voltaire born.
1698		Savery invented a mine pump.	
1699			Royal College of Art established at Berlin.
1700–1800		European nations fashioned their policies according to the mercantile theory.	
1700			Yale College founded.
1704			John Locke died.
1705		Neucomen perfected the mine pump.	
1712			J. J. Rousseau born.

THE ADVANCE OF VOCATIONAL EDUCATION

Year	Political-Legislative	Economic-Social	Educational-Philosophical
1719		Irish linen makers set up a loom at Londonderry, New Hampshire.	
1734	Washington born.		
1738		Key invented the flying shuttle.	
1739			Sunday School made compulsory, Wurtemberg, Germany.
1740	Frederick the Great became King of Prussia.		
1747			Ecole des Ponts et Chaussées opened in Paris.
1747			Hecker-Realschule opened.
1755			Society of Arts founded in London.
1760	George the Third, King of England.		
1762			Rousseau published "Emile."
1763	End of the Seven Years War.		
1764	The Sugar Act passed.		
1765	The Stamp Act passed.		
1765			Royal Saxon School of Mines opened at Freiberg.
1766			Oberlin began work at Waldbach, Germany.
1767	Townshend Acts passed.		
1767		Hargreaves invented the spinning jenny, which improved hand spinning.	
1768			Royal Academy opened in London.
1769		Watt's improvements in the steam engine gave it use in the factories.	
1769		Daniel Boone began western explorations.	
1769	Napoleon born.		
1770	Boston Massacre.		
1771		Arkwright invented the water frame for spinning.	

THE ADVANCE OF VOCATIONAL EDUCATION

Year	Political-Legislative	Economic-Social	Educational-Philosophical
1773	Boston Tea Party.		
1774			Basedow's Philanthropium founded.
1774			Kinderman organized the "Schools of Industry" in Bohemia.
1776			Adam Smith advocated education of apprentices in England.
1776	Declaration of Independence.		
1778			Voltaire died.
1778	Clark seized French villages along the Ohio River.		
1779		Crompton invented the spinning mule.	
1783		Treaty established Mississippi River as western boundary of the United States.	
1785		Cartwright invented the power loom.	
1786	Frederick the Great of Prussia died.		
1787			Board of Regents, University of the State of New York, founded to coordinate all secondary and higher educational activities of the state.
1787	Slavery prohibited in the Northwest Territory.		
1787		First cotton factory in the United States founded at Beverly, Mass.	
1788			English Philanthropic School established "Industrial Reform School."
1788	Constitution of the United States ratified.		
1789			Jefferson proposed a scheme of universal education based upon the theory that a modicum of education for all and special training for the more capable should be offered at public cost.
1789	First tax laid by Congress for Federal purposes (tax on imports).		

Year	Political-Legislative	Economic-Social	Educational-Philosophical
1790	Benjamin Franklin died.	Slater set up English spinning machinery in Rhode Island.	
1790		Adam Smith died.	
1791			"Schools of Industry" established in England.
1791		Alexander Hamilton submitted a report to Congress on the industries of the country.	
1792			Thomas Paine published "Essay on Rights of Man."
1792	Congress established a system of coinage.		
1793		Eli Whitney invented cotton gin.	
1794			Compulsory school attendance in Germany.
1795			Ecole Polytechnique opened in Paris.
1796	Report of the Manchester (England) Board of Health.		
1798			Pestalozzi opened orphan school at Stanz.
1798			Adult school at Nottingham (England) opened.
1799			Robert Owen began social reforms at New Lanark.
1799			First National School of Art and Trades opened in Paris.
1799			Niemeyer advocated teaching of the manual arts.
1799			Fellenberg purchased Hofwyl.
1800			Birkbeck—beginning of Mechanics' Institute Movement in England.
1802	First Factory Act passed in England.		
1802		Cumberland road begun as a national highway.	
1803			Sunday school made compulsory in Bavaria.
1803			Emerson born.

THE ADVANCE OF VOCATIONAL EDUCATION

Year	Political-Legislative	Economic-Social	Educational-Philosophical
1803	Louisiana purchased.		
1803		New York Society of Journeymen Shipwrights formed.	
1804–1824			Pestalozzi—Institute at Yverdun, Switzerland.
1804	Alexander Hamilton died.		
1804			Kant died.
1805			Pennsylvania Academy of Fine Arts founded at Philadelphia.
1806		New York Carpenters organized.	
1807		Fulton applied steam to navigation.	
1807	Embargo and non-intercourse acts passed.		
1807			Fellenberg Academy built at Hofwyl.
1807			Fellenberg farm and trade school under Wehrli.
1807			Boston Athenaeum founded.
1808			American Academy of Fine Arts founded in New York City.
1812	War with England.		
1814	Abdication of Bonaparte.		
1814		Locomotive invented.	
1814		Francis Lowell perfected a power loom at Waltham, Mass.	
1814			Farm and trade school established in Boston.
1814			Compulsory apprenticeship abolished in England.
1815	Bismarck born.		
1816			Owen opened infant school at New Lanark.
1816	Congress passed a protective tariff.		
1816		Regular steamboat navigation of western rivers began.	

Year	Political-Legislative	Economic-Social	Educational-Philosophical
1817		Puddling process for refining iron introduced into United States from England.	
1818			Falk opened "Industrial School for Orphans" at Weimar, Germany.
1819			Pounds established ragged school at Portsmouth, England.
1820		From this date a steady rise in the output of the manufacturing industries of the United States can be noted.	"Mechanics School" opened in New York City.
1821	Florida acquired from Spain.		
1821			Fowle taught drawing in a Boston public school.
1823			Gardiner Lyceum opened in Boston.
1823	President Monroe formulated the "Monroe Doctrine."		
1824			London Mechanics' Institute opened.
1824			New York House of Refuge opened.
1824			Franklin Institute founded in Philadelphia.
1824			Rensselaer Polytechnic Institute founded at Troy, N. Y.
1824	Congress passed a protective tariff.		
1825		Erie Canal opened.	Owen opened school at New Harmony, Ind.
1825			Manual labor at Maine Wesleyan Seminary.
1826			Froebel wrote "The Education of Man."
1826			Franklin Institute opened high school at Philadelphia, Penna.
1827		Philadelphia carpenters struck for a ten-hour day.	
1827			Fellenberg began work at Meykirch.

THE ADVANCE OF VOCATIONAL EDUCATION

Year	Political-Legislative	Economic-Social	Educational-Philosophical
1827			Oneida Institute opened at Whitesborough, N. Y.
1828			Ohio Mechanics' Institute opened in Cincinnati.
1829			Ecole Centrale des Arts et Manufactures founded in Paris.
1830			Woodbridge published "Sketches of Hofwyl."
1830–1840		Anthracite substituted for charcoal in the smelting of iron.	
1830	House of Commons appointed a committee on art education.		
1831		Power applied to furniture making.	
1831			Society for Promoting Manual Labor founded in New York.
1831		Baltimore and Ohio Railroad operated steam-locomotive train.	
1832	Jackson vetoed bill to continue the United States Bank.	Economic result the "Wild Cat" banks.	
1832		Local organizations in New York formed the general trades union of New York City.	
1832		Power knitting mills were set up at Cohoes, N. Y.	
1833		Patent taken out by McCormick on grain reaper.	
1833			First grant for elementary schools by English Parliament.
1833	Factory Act, passed in England, provided half-time schools for factory children.		
1833			Wehrli became principal of the normal school at Kreuitzlinger, Switzerland.
1834			Alcott opened the Temple School, Boston.
1835			First "Civil Engineer" graduated at Rensselaer Polytechnic Institute.

264 THE ADVANCE OF VOCATIONAL EDUCATION

Year	Political-Legislative	Economic-Social	Educational-Philosophical
1835			New England school districts had become autonomous.
1835–1870			Era of public school revival in the United States.
1835			The heydey of the private academy; the monitorial school was the chief type of training institution.
1836–1838			Bache studied schools in Europe.
1837	Financial panic in the United States.		
1837			Massachusetts established a State Board of Education with Horace Mann as secretary.
1838	Victoria became Queen of England.		
1839		Typographical Society of New Orleans, La., adopted a provision requiring apprenticeship for a term of not less than four years during minority.	
1839			Kay discussed pauper schools in England.
1839	First laws in Germany relating to factory children.		Schools for factory children considered in Germany.
1840		Brass substituted for wood in clock manufacture.	
1841	France passed a law regulating the labor of factory children.		The laws of France regulating child labor made way for schools for these children.
1842			Bayley founded "Peoples' College" at Sheffield, England.
1842			William James born (died 1910).
1842	Massachusetts Supreme Court decided that labor organizations were legal.		
1843		Rothamsted Experimental Farm established at Harpenden, England.	
1844			Ashley formed the Ragged School Union, England.

THE ADVANCE OF VOCATIONAL EDUCATION

Year	Political-Legislative	Economic-Social	Educational-Philosophical
1844		Paris and Orleans Railway Company adopted a profit-sharing scheme.	
1844			New York established the Albany State Normal School, David Page, principal.
1845		Workingman's Protective Union founded, Boston.	
1845	Potato crop failed in Ireland.		
1846		Howe invented the sewing machine.	
1846	War between United States and Mexico.		G. Stanley Hall born.
1848	War in Germany.		
1848, 1849		Gold discovered in California. Rush to California, migration of western states, and the concomitant economic and financial reactions.	
1848			Germany established free public elementary schools.
1848			Girard College for orphans opened in Philadelphia, Penna.
1850–1880		Era of railway development.	
1850		Coal used as locomotive fuel in United States.	
1850		Waltham (Mass.) company started to manufacture delicate watch mechanisms.	
1850		Specialization noted in uni-product factories.	
1850	English government received report on needle industries.		
1850			Turner delivered an address on an industrial university.
1851		World's Fair, London.	
1852			Training teachers became the main purpose of the English government's school of design.

266 THE ADVANCE OF VOCATIONAL EDUCATION

Year	Political-Legislative	Economic-Social	Educational-Philosophical
1852			Turner proposed national land grants for universities in America.
1853			Antioch College founded by Horace Mann.
1853	Germany passed law regulating child labor in factories.		
1854			Maurice opened Working Men's College in London.
1854			Ruskin taught drawing in Working Men's College in London, England.
1855			Royal School of Art and Industry opened in Munich.
1855–1881			Henry Barnard published the Journal of Education, 31 volumes.
1856		Bessemer invented process for making steel rapidly and cheaply, England.	
1857	Congress passed lower tariff law.		
1858		Blake invented machine for sewing shoe uppers to soles.	
1859			Norris (England) discussed industrial training for girls.
1859			Darwin published "Origin of Species."
1859			John Dewey born.
1860		Sewing machine applied to boot and shoe industry.	
1861–1870		Immigration into United States for this period 151,824.	
1861	Civil War began.		
1861–1865		Steel manufacture perfected.	
1861			Herbert Spencer published "Education," four essays—"The aim of education is complete living —which consists in physical well-being; vocational capacity, parenthood, citizenship, and appreciation of the finer things in life."

THE ADVANCE OF VOCATIONAL EDUCATION

Year	Political-Legislative	Economic-Social	Educational-Philosophical
1861		Beginnings of standardization noted—first in manufacture of fire arms.	
1861		First oil "gusher" struck.	
1861–1890		Discoveries and examinations of mineral resources in United States.	
1862	Homestead Act passed.		
1862	Morrill Act passed providing for "land grant" colleges.		
1862	United States Department of Agriculture established.		
1862		Union Pacific Railroad Company created.	
1862			Charles R. Allen born.
1864–1864	National Banking Acts passed.		
1864		Bessemer process introduced to America—Michigan.	
1864			James Earl Russell born.
1864	Congress passed the Alien Contract Immigration law.		
1865	Civil War ended.		
1865	Lincoln was assassinated, Johnson, President.		
1865			Report of the Commission on Technical Education submitted, France.
1865			French commission discussed shop work and advocated drawing.
1866		Atlantic cable laid.	
1867		Abram Hewitt (United States) introduced open hearth process for reducing iron ores.	
1867			Henry Barnard became first U. S. Commissioner of Education.
1867		Seamless knitting machine invented.	
1867	Massachusetts prohibits employment of children under 10 years in manufacturing.		

268 THE ADVANCE OF VOCATIONAL EDUCATION

Year	Political-Legislative	Economic-Social	Educational-Philosophical
1867	Alaska purchased.		
1867		National Grange formed.	
1867	Wines surveyed juvenile reformatories.		
1869			Washburn shops opened at the new technical school at Worcester, Mass.
1868			Hampton Institute for negroes opened by Armstrong.
1868			David Snedden born.
1869		Knights of Labor formed.	
1869	Germany passed "Regulation of Industry," making continuation school attendance compulsory.		
1869		Union Pacific Railroad opened.	
1869	Massachusetts towns were authorized to establish evening schools.		
1870–1930		Tendency toward concentration of money, men, machines, and management noted in American economic life.	
1870	Franco-Prussian War.		
1870	England adopts compulsory school attendance.		
1870			Metropolitan Museum of Art founded, New York City.
1871			Charles Prosser born
1872		Refrigerator cars first used.	John W. Withers born.
1872			Herman Schneider born.
1873	Length of Massachusetts school year extended to 20 weeks. Compulsory age attendance raised to 12 years.		
1874			Lake Shore & Michigan Central R. R. began an apprentice school.
1876		Bell perfected the telephone.	
1880		Public scrutiny of the conduct of business began.	

THE ADVANCE OF VOCATIONAL EDUCATION

Year	Political-Legislative	Economic-Social	Educational-Philosophical
1880–1928	Laws regulating ventilating, lighting, and sanitary arrangements in manufacturing plants passed by most of the industrial states.		
1880		Number of females gainfully employed, 2,647,157 (United States census).	
1881		Standard Oil trust formed.	
1881		Period of industrial amalgamation and integration began.	
1881		American Federation of Labor founded.	
1881			Auchmuty founded the New York Trade School.
1882	Immigration for year was 788,992.	Labor market glutted with cheap European labor.	
1883	Massachusetts towns compelled to establish eveningschools. (Population minimum set at 10,000).		
1885	Congress passed law prohibiting companies prepaying transportation of aliens.		
1886	Senate report on railroad discriminations presented.		
1887	Hatch Act passed.		
1887	Interstate Commerce Act passed.		
1888	Children under 13 excluded from Massachusetts factories.		
1889	Compulsory school age raised to 14 in Massachusetts.		
1890			James published "Principles of Psychology."
1890			William Russell born.
1890	Sherman Anti-trust Act passed.		
1891			Baron de Hirsch Trade School organized.
1893		I. W. W. founded.	

270 THE ADVANCE OF VOCATIONAL EDUCATION

Year	Political-Legislative	Economic-Social	Educational-Philosophical
1894		Northrop Automatic loom perfected.	
1894	Erdmann Act passed providing for mediation and arbitration of industrial disputes.		
1894	Minimum-wage law in New Zealand.		
1895		National Association of Manufacturers formed.	
1897		Illinois Coal Operators' Association formed.	
1898			James Earl Russell named Dean of Teachers College, Columbia.
1899		The National Metal Trades Association formed.	
1899			First public school classes for crippled children in the United States conducted in Chicago.
1901–1910		Immigration for period—8,795,386.	
1901	Queen Victoria died.		
1901			Thorndike made professor of educational psychology, Teachers College, Columbia.
1902		Wireless telegraph perfected.	
1902	Congress passed the Reclamation Act.		
1902		Anthracite Coal Strike.	
1903	Congress established Department of Commerce and Labor.		
1904			Hall published "Adolescence."
1904			John Dewey made professor, Columbia University.
1904		The Barber warp-tying machine used in the cotton industry.	
1904		The Nernst Lamp Company of Pittsburgh introduced a "factory committee" to give employee representation on shop councils.	

THE ADVANCE OF VOCATIONAL EDUCATION

Year	Political-Legislative	Economic-Social	Educational-Philosophical
1906			Snedden, State Commissioner of Education, Massachusetts 1906–1916.
1906			New York City opened classes for crippled children.
1906			Herman Schneider made dean, University of Cincinnati.
1906			University of Wisconsin established the university extension division.
1906			Report of the Massachusetts committee on Industrial and Technical Education.
1906		Exhibit of safety devices and appliances held in New York under the auspices of the New York Institute for Social Service.	
1907			City of Milwaukee, Wis., took over the Milwaukee school of trades.
1908			Dean Schneider developed the cooperative plan of engineering education at the University of Cincinnati.
1909–1911			Era of establishment of public industrial secondary schools in large cities of U. S.
1909			Boston Y.M.C.A. started cooperative night school work.
1909			Sante Fe Railway inaugurated an apprentice training system.
1909	Payne-Aldrich Tariff passed.		
1910		Number of females gainfully employed was 8,075,000. (United States census).	
1911			Attendance at part-time classes made compulsory for employed boys and girls in Wisconsin.
1911			Pennsylvania Railroad began its apprentice school.

272 THE ADVANCE OF VOCATIONAL EDUCATION

Year	Political-Legislative	Economic-Social	Educational-Philosophical
1911		The Supreme Court decided for dissolution of the Standard Oil Company and the American Tobacco Company.	
1911–1920		Immigration figure for the period 5,735,811.	
1911			David Snedden published "Problems of Vocational Education."
1912		National Safety Council founded.	
1913	Federal Reserve Act passed.		
1913	Underwood Tariff passed.		
1914	Federal Trade Commission established.		
1914	World War began.		Thorndike published "Psychology of Learning."
1914		First ship sailed through the Panama Canal.	
1914			Report of the Commission on National Aid to Vocational Education.
1915		The Colorado Fuel and Iron Company introduced a plan whereby employees were to be elected to act on committees functioning on matters pertaining to sanitation, recreation, wages, education, and employment.	
1915	Pennsylvania passed an act providing for continuation schools.		
1916	United States Shipping Board created.		
1916			John Dewey published "Democracy and Education."
1917	United States entered the World War.		
1917	War Industries Board founded.		
1917	Congress passed an act refusing admission to all aliens over sixteen, physically capable of reading, who could not read English or some other language.		

Year	Political-Legislative	Economic-Social	Educational-Philosophical
1917	Wisconsin passed a law making attendance in a part-time school compulsory for all boys and girls, not high school graduates, up to 18 years of age.		
1917	Smith-Hughes Act passed.		
1917	Federal Board for Vocational Education created.		
1918	Fisher Bill for compulsory continued education passed by Parliament.		
1918			Carnegie Foundation for the Advancement of Teaching issued a report on Engineering Education.
1918		United States government took over the railroads.	
1918	Armistice signed.		
1918		National Association of Employment Managers organized.	
1919		The steel strike occurred in the United States.	
1920	Congress passed the Civilian Rehabilitation Act, providing for the vocational rehabilitation of sane persons.	8,500,000 females gainfully employed in the United States. (United States census.)	
1920		The American railroads were returned to private ownership.	
1921	Congress passed an act limiting the annual immigration from any one country to three per cent of the number of foreign born residents of that nation enumerated in the census of 1910.		
1921			John W. Withers made Dean of New York University, School of Education.
1921			New York University offered cooperative courses in engineering.
1921			Snedden published "Sociological Determination of Objectives in Education."

Year	Political-Legislative	Economic-Social	Educational-Philosophical
1922			William Russell published "Trend in American Education."
1922		Coal strike in the United States.	
1922	Fordney-McCumber Tariff passed.		
1923	Coolidge made President of the United States.		
1924	Congress extended the appropriations for the Rehabilitation Act for six years. Congress passed an act extending the benefits of the Smith-Hughes Act to the territory of Hawaii.		Department of Vocational Education established at School of Education, New York University.
1925		Radio perfected.	
1927			William F. Russell, dean of Teachers College.
1929	Herbert Hoover inaugurated President of the United States.		
1929	United States Farm Relief Board appointed.		
1929	The law limiting the immigration of aliens to the United States in the ratio of "national origins" went into effect.		
1929	Congress passed the George Reid Act, providing additional sums of money, amounting to $2,500,000, for vocational agriculture and vocational home economics.	Excess speculation in stocks which commenced in 1927 resulted in break in stock prices on the New York Stock Exchange.	
1929	Congress passed an act extending the benefits of civilian vocational rehabilitation to the District of Columbia.	President Hoover called into conference the leaders in industry and business of the country in order to strengthen the economic foundations of trade and industry.	
1929	United States became a member of the World Court. The Pact of Paris signed. (The Kellogg-Briand pact to outlaw war).		
1930	The Smoot-Hawley Tariff passed by Congress. The London Naval Treaty passed the United States Senate.	Stock market maintained its deflated condition. Drought ruined the crops of numerous farmers throughout the United States. Industry suffered from the wave of depression. Unemployment high.	
1930	Congress extended appropriations for civilian vocational rehabilitation for three years.		

Part Eight

THE ROOTS OF VOCATIONAL EDUCATION

A SUMMARY

THE ROOTS OF VOCATIONAL EDUCATION

This section is an attempt to recapitulate the more important movements that have contributed to vocational education. In the previous section, "Milestones in the Advance of Vocational Education," stress was laid upon the minute events which, when integrated, have formed the structure of vocational education as it exists in twentieth century America. It now becomes necessary to review the major trends in the psycho-social environment that might be isolated and termed roots, from which vocational training schemes can be traced. These roots are fixed in the changes which have taken place in the social, political, and economic life of the nation. It is hoped that, when the various roots have been enumerated, the outgrowing agencies of training, which are so closely connected with these roots, will stand forth in clear-cut fashion. This section makes no attempt to repeat the lengthy discussions of the foregoing sections, but rather it is offered as an epitome and recapitulation of the whole study.

The English Background of American History.—No attempt is made here to restate the contributions of the pre-Christian era, the middle ages, or the years before the discovery and colonization of the American continent. However, it is important to review life in England at the time when the colonists set forth for the American shores. England, at that period, was an agrarian nation. Its people were, for the most part, engaged in farming and sheep raising. What little manufacturing there might have been, was done by hand in the home or shop. The customs and traditions of the medieval guilds flavored this domestic system of manufacture. Education of the formal sort was for the nobility, the clergy, and the rich. These groups constituted the upper strata of a society in which the dualism of labor and leisure was extremely marked. From this scene the Pilgrims embarked for western shores. This social heritage was a part of them when they set out for the new world.

Agriculture, the Basis of Colonial Life.—When they landed in America, they found a virgin country covered with forests and possessing no established culture. It is obvious why agriculture became their first great occupation: the need for food, shelter, and clothing was primary. As a result farms took the place of the wilderness. The agricultural influence, after a number of years, worked its way into the schools, and the early academies gave place to agricultural subjects and training in their curricula. The Reverend Mr. Monteith's Manual Labor Academy at Germantown, Penna., is a sample of a school which attempted to provide agricultural training. Other examples include the Oneida Institute of Science and Industry and the Robert Owen venture at New Harmony, Ind. (See Part Two.)

The Industrial Revolution in England.—While the early colonists were struggling with the sterile soil and sandy lands along the North Atlantic seaboard, a gigantic upheaval was taking place in England. The Industrial Revolution occurred in England after the year 1750. Enough has been said in Part One of this work about the causes leading up to this great and sweeping change. The Industrial Revolution resulted in the substitution of power-driven machinery for hand labor. With it came the factory system of manufacture, which resulted in the modern industrial plant, the urbanization of population, the growth of labor organizations, and the development of scientific research in industry. Social, political, and economic life were revolutionized by this change in the method of manufacture. The modern world is largely a product of this momentous movement.

The Industrial Revolution Delayed in America.—Until after the War of 1812, however, the Industrial Revolution did not manifest itself in America. England, for many reasons, endeavored to use the western world as a source of raw materials from which to supply her factories and as a market in which to sell her manufactured products. After the second war with England, the Americans realized that they must provide themselves with the means of large-scale production. The government was not the last agency to assist in the upbuilding of American industry, and the series of protective tariff laws did much to stimulate home production. After 1820 manufacturing sped forward in America by leaps and bounds. The various statistics given in Part One tell the story of this evolution in great detail. Of course it is to be remembered that the great natural wealth of the country was a fundamental basis upon which an industrial order could be firmly established. The indomitable spirit of the American people was another item not to be neglected.

It may be said, however, that after 1850 America became an industrial nation. By the close of the nineteenth century, America was preeminently the industrial nation of the world. The population had become urbanized, and the large industrial plant was the distinctive feature of the American city.

The Factory System a Root of Vocational Education.—The modern factory system is a major root from which vocational education has developed. In order to connect the various agencies of vocational training with their roots, it is better to subdivide this mighty root represented by the Industrial Revolution. The phases of the Industrial Revolution which can be considered of major import include (1) the use of power driven machinery in industry, (2) the division of labor, (3) the shortening of the hours of labor, (4) the increase in per capita wages, (5) the entrance of women into factory labor, (6) the development of scientific research in industry, and (7) the problem of factory management and administration.

The Change from Hand Labor to Machine Production Resulted in a Change of Training.—Perhaps the most evident change in industry caused by the Industrial Revolution was the change from hand production to machine production. The great inventions of the eighteenth and nineteenth centuries made possible the mechanization of industry. This change required a change in the training of workers for positions in industry. All but a very few of the schools of the early part of the second half of the nineteenth century dealt only with the cold and rigid exercises designed for mental training, and did nothing in the way of providing future industrial workers with experiences that might prove helpful to them in the tasks and problems of everyday life. A few academies did provide woodworking shops, but these were exceptions, and the courses of study in these places stressed wood-working as a form of training useful for mental transfer. To meet the needs of the new industries, the private trade schools arose. Many were founded after 1880. These include the New York Trade School (1881), the Philadelphia Builders' Exchange (1890), and the Baron de Hirsch Trade School (1891). (See Part Two.)

The Apprenticeship System Failed to Provide Efficient Workers in Sufficient Numbers.—Up to this period the apprenticeship system was looked to as the means whereby industry might be provided with capable workers. After the passing of leisurely craftsmanship and the substitution of the machine worker, however, it became evident that the apprenticeship system, with its prolonged training period, was inadequate to meet the increasing needs of a progressive, industrial nation. The private trade schools arose to answer the demands for trained workers.

The Division of Labor as a Root of Vocational Education. The present industrial order is characterized by a minute division of labor. This is due to a variety of causes among which stand out the following: (1) The intricacies of modern machinery preclude the possibility of the worker mastering the techniques of operating many machines or performing many operations; (2) the efficient factory lay-out and the efficient flow of production demand a segregation of operations and machines of the one sort; and (3) greater efficiency may be obtained by a worker who is practiced in a given operation or on a given machine.

Herein lies another root from which may be traced vocational training schemes. In order to provide operators skilled in the use and knowledge of complex and highly specialized machinery, the factories themselves established training schools.

The vestibule schools, the foremanship training programs, and the corporation schools are offshoots of this root. Part Four of this work is devoted entirely to the efforts of business and industrial firms to prepare their workers for efficient service. The schools of the General Electric Company, the Westinghouse Electric Company, the New York Central Railroad, and the Yale and Towne Manufacturing Company are concrete evidence of the fact that industry itself has been vitally interested in vocational training.

The Shortened Hours of Labor, a Root of Vocational Education.—The shortening of the hours of labor constitutes a further root from which vocational education can be said to have sprung. With the reduction of hours from fourteen to eight per day it became evident that workers could engage in further education during their spare time. The evening school was the agency which first attempted to provide education for people engaged in industry. As the hours of labor were curtailed, the continuation school came forth as the agency in which adolescents engaged in labor could find an opportunity for study during the day. The early evening schools include, among others, those conducted by various cities, the Young Men's Christian Association, and by the factories themselves. The early evening schools attempted to carry on the general education of their students, while those of today have widened their scope so as to include, in addition, specialized courses of a vocational character.

The Continuation School, a Result of Shorter Hours of Labor.—In the continuation school field, the story is quite similar. The general continuation school type of institution prevailed at first. This type sought to give general education to its students; it was a continuation school in the literal sense of the word. Later, due to the insistent demand of industry for adequately trained workers, a second type of continuation school, the trade-extension school, arose. The purpose of this school was to provide training in the processes of production which would improve the performance of the workers in the shop. The Central Needle Trades School and the Central Printing Trades School of New York City are examples of this type of school.

At present there is a movement for continuation schools which will be opened all day for workers who may wish to come to learn particular jobs or specific operations. The work of the Frank Wiggins Trade School in Los Angeles, although this is not a continuation school, and the Denver Opportunity School are samples of this "open-type" trade extension education which makes provision for "pusher" education. Public continuation schools might well copy the activities of these schools.

Higher Wages Have Made Possible Continued Education. A fourth feature of the present industrial system is the higher wage paid to industrial workers. The added salary is a means by which individual workers may seek further education. The New School for Social Research in New York City, the trade extension schools, and the evening and part-time engineering colleges throughout the country are institutions which cater to the needs of the more affluent workers. The desire which many workers have for promotion to the semi-executive and executive positions in the factory is the drive which has made possible the correspondence school. The International Correspondence School, of course, is a typical example of a school made possible by the patronage of well-paid workers who have some time for study.

Women in Industry, a Result of Machine-Production.—The entrance of women into industry has doubled the whole problem of industrial training. The ease with which automatic machines can be tended, the short hours of labor, and the tempting wage have been instrumental factors in luring women into the factories. The factory schools, the trade schools, and the continuation schools have broadened their scopes so as to include facilities for the training of women. The Manhattan Trade School for Girls and the Boston Trade School for Girls, among others, are the result of women's advent into industry.

The Research Worker and His Function in Modern Industry.—Scientific research in industry has created the demand for research workers both of college and sub-college grade for positions in industry. The birth of modern science can be traced back to the rationalistic era which occurred about the time of the French Revolution. After that time, however, science took a new turn. The influence was away from pure and theoretical studies, and toward the fields of engineering and applied science. The American colleges were not slow in seeing that the directive agents for the industrial world must be technically trained experts. After the third quarter of the nineteenth century engineering schools grew up throughout the country. Massachusetts Institute of Technology, the School of Applied Science (now the College of Engineering) at New York University and Cooper Union Institute are notable examples of this development. For research workers of less than college grades, junior research workers so-called, the factory schools have assumed the major part of the burden. The General Electric School is a fine example of this type of institution. However, there have been established numerous public technical schools of the secondary grade such as the Brooklyn (New York) Technical High School in New York City, the Utica (New York) Free High School, and the Dickinson High School in Jersey City, New Jersey.

Industrial and Administrative Engineering.—The problem of factory management and administration has been cared for largely by the colleges of engineering of the United States in their departments of industrial engineering. The University of Cincinnati, the New York University and the Georgia Technical Institute are examples of engineering schools which have attempted to supply industry's demand for leaders of ability and training. The schools mentioned realized the shortcomings of theoretical training and have carried on their work by means of the cooperative system of training. The students spend their time between the school and the shops, studying under college conditions, and working on real jobs in a true factory atmosphere. (This type of school has been explained in detail in previous sections of this treatise.)

Modern Industrialism, the Basic Root of Vocational Education.—These characteristics of modern industrialism are all roots of the present system of vocational education. In each case it has been seen that from these roots, the inherent characteristics of the factory age have blossomed into actual agencies, that is schools of one sort or another, which have had the training of workers as their common goal. Briefly, a need arose in industry, and society provided for its satisfaction by evolving an agency equipped to provide for the need. Many times, as has been pointed out in the bulk of this study, the agency itself has been imperfect, inadequate, and often of only temporary assistance. This, however, is remote from the point. Vocational education has arisen in answer to definite demands, whose roots are inextricably bound up in the present factory system of production as established by that sweeping change in techniques and procedures known as the Industrial Revolution. The roots enumerated above are merely suggestive of the great mass of roots provided by the factory system. It is to be remembered that the influences of these roots were felt very promptly, and that they acted together and not separately. The complexity of the problem has been due largely to this rapid and concerted action.

The Scientific Movement in Education.—Modern science was not slow in creeping into the experiments and studies of modern educators. After the dawn of the twentieth century, the science of education took on a new and more practical aspect. The work of James, Hall, Thorndike and others has been mentioned. The significant studies in school populations made by recent investigators have been discussed at length in Part Three. It is necessary here to mention only the results of these studies in general terms. The fact was clearly shown that the greater part of the elementary school population entered industry rather than the colleges. The studies of Van Denburg, Ayres, and Counts gave ample evidence of the failure of the schools to hold the pupils and to provide for those who were to leave at an early age. As a result of these and other studies, the schools adapted themselves as best they could to the situation. A large proportion of the students were to enter industry either before or after they had completed their secondary school work. The present system of all day vocational schools and the elaborate scheme of industrial arts education were developments arising from this specific root of vocational education. The Boys Vocational School and the Brooklyn Technical High School were attempts to provide for youthful entrants into industry. Other examples could be given to include the Textile High School of New York City and the hosts of technical high schools throughout the country. However, the elimination studies of the educators brought to light a further fact, that many students left school years before they entered the high school. Herein lies the root from which the industrial arts education sprang. This type of education was planned to provide some industrial training for the many children whose days in the school were limited to the pre-high school period.

The industrial arts training was assisted by another movement provided for by the modern educators. The experiments of the educational psychologists blasted the theory of the transfer of training. This resulted in the abandonment of the courses in manual training. Manual training had been held up as a means by which a carry-over to mind training from hand training might be established. With the foundation shown faulty, the manual training schemes were transformed into industrial arts courses.

The Problem of the Destitute Child Led to Early Attempts at Trade Training.—The urbanization of population brought about by the factories and by the consequent localization of production centers caused many social problems. One of the first of these problems that aroused the interest of society was that of the care of the destitute child. Part Two of this work gives in detail the story of the early attempts at the solution of this most intricate problem. The homes for these unfortunates were, from the earliest days, centers of vocational training. It is not always true that this training was given with the altruistic motives which should have been the guiding force of these institutions. However, the homes for poor and delinquent children constitute a root from which vocational education received an impetus. Examples of these institutions include such noted establishments as the New York City House of Refuge and the public truant, probationary, and parental schools, especially those of New York City.

The Fraternal Societies Have Aided Vocational Education.—Closely allied to these homes established by society for unfortunate children stand the educational facilities provided by the fraternal orders for the children of their members. The Masons and the Loyal Order of Moose have been leaders in the field of providing vocational training. In a list of the roots of vocational education the work of these associations is not to be omitted. Mooseheart, mentioned in Part Three, is a complete town devoted to childhood, and it is interesting to note that vocational education is provided under capable directors. It has been the aim of the schools provided by these associations to enable their graduates to find places in the business or industrial world, where they might become successful and happy citizens.

The American Federation of Labor and Vocational Education.—Another social agency which has done much in stimulating and backing vocational education is the American Federation of Labor. Organized labor was a natural outgrowth of the massing of men together in factories. The guilds were the forerunners of the modern labor unions. After the establishment of the factory system, it was not long before the workers banded together for mutual aid. The history of the rise of labor organizations in America has been treated in the foregoing sections of this study. From its foundation in 1881, the American Federation of Labor has consistently advocated vocational education in its public utterances. Convention after convention of this group has expressed sentiments favoring the various types of vocational training. Moreover, the Federation has assisted notable educational surveys, the Richmond survey being a typical case in point. The Milwaukee Trade Schools are further evidence of the attitude of cooperation shown by organized labor. Truly, organized labor is an important root of the present day set-up of vocational training.

The pamphlet, "Education for All," is an expression of

labor's attitude on questions of vocational training. Ample quotations therefrom are given in the section of this work entitled "Labor and Vocational Education." Samuel Gompers and other great labor leaders have been moving figures in the vocational education crusade.

Liberalism and the Spirit of Democracy, Basic Roots of Vocational Education.—Another, and somewhat intangible, root of vocational education has been the general growth of a liberal attitude on the question of what constitutes an education. The results of the studies in psychology and sociology of the modern age have resulted in creating in the general public a realization that Latin and Greek are no longer the sole distinguishing marks of the cultured man, but that real culture consists of the optimum use of one's energies and abilities, enhanced by adequate training for efficient work in the right occupation, and of the intelligent disposition of one's leisure time. This liberal attitude has done much to remove the social stigma formerly attached to the students and even the faculties of the vocational schools. In the old days, vocational schools were apt to be considered as schools for wicked and incorrigible children, supervised by uncultured and uncouth people who took upon themselves the sacred title of "teacher." Today, it appears that much of this complacent feeling of superiority on the part of academic teachers has given way to a feeling of respect and to a recognition of those whose work is somewhat different and whose training is of the shop rather than of the seminar.

The Modern Notion of the Function of the State.—A movement of recent date has concerned itself with the question of the rights and responsibilities of the State. The liberal attitude of some of the more progressive governmental leaders has resulted in a series of laws which aid education, and especially education of a vocational sort. Legislation has proved another vigorous root to which vocational education may be attributed. In the early days of the country's history, farming constituted the major industry of the people. The farms were small in size, and the methods employed were direct carry-overs from those of Europe. As the great West was opened to settlers, farms increased in size, and specialization of crops became a new feature of American farming. The small farmer of 1790 raised a variety of products; the large farmer of 1910 specialized in a single crop.

After the Civil War, the Industrial Revolution spread to farming. The mechanical reaper of Cyrus McCormick is but one of the many inventions which improved farming activities. Closely following the numerous inventions of mechanical farm implements there came a movement resulting in the application of chemistry and geology to farming. "Scientific farming" arose as a distinctive feature of modern American agriculture.

The present-day land owner in the farming areas of this country finds it necessary to be technically trained in the science of his occupation. In order to provide such training, New York State established an agricultural school at Cornell University. The schools at Syracuse and the innumerable agricultural high schools and agricultural stations throughout the farm areas are vocational training agencies whose origins were tied up in the development of farming and the applied science of agriculture.

Legislation Has Given Formal and Material Support to Vocational Education.—Another, and perhaps the most interesting, root of vocational education has been legislation. The evolution of the modern state and the definition of its duties and powers is a phase of social history that has often been overlooked. In the realization that farming was to be one of the chief occupations that was to engage the people of the western states, Congress in 1862 passed the first of a series of laws whose purpose was to aid scientific farming by assisting land-grant colleges of agriculture. The famous Morrill Act was a monumental root from which other acts favorable to vocational education can be traced. A previous section of this study recounts the history of Federal aid to vocational education. After the passage of the Morrill Act, the great state universities of the West grew up. These have proved themselves to be foundation stones of American education.

FEDERAL AID TO VOCATIONAL EDUCATION

The interest of society in the child has tempered all social legislation. Especially is this true of legislation dealing with education. The realization that the health and happiness of children is a fundamental necessity for the security of the future state has led to legislation protecting the child. Laws regulating his employment are on the books of all the states. The second step in providing for the child's well being was the compulsory school laws. As the years have passed, the age at which children are permitted to leave school has been raised.

As a result the school has been forced to provide means for taking care of these great hordes of children who are by law compelled to remain in school. The all-day vocational schools and the industrial arts courses have been designed to meet this problem. For those who must go to work at an early age, the continuation school has been developed, and from such children attendance in school is required but a few hours a week. Compulsory school attendance laws, then, also constitute a root of vocational education. The broadened curricula has been the result of the changed character of the school population due to the operation of the compulsory attendance laws.

Federal Aid to Vocational Education.—The agitation for Federal aid for vocational education was based upon the principle that the continuation of the present industrial system is fundamental to the future strength and happiness of the country. "Vocational Education in a Democracy," by Prosser and Allen, is a book in which this phase of the evolution of vocational training is convincingly portrayed. Many agencies joined in the demand for Federal aid. To mention the most prominent campaigners one must include industry, labor, the engineering societies, the fraternal associations, educators, parents, industrial workers, and social workers. The result was the passage in 1917 of the Smith-Hughes Act and the creation of the Federal Board for Vocational Education. The results of this epoch-making piece of legislation have been the founding and improving of countless vocational schools. The 1930 report to Congress of the Federal Board for Vocational Education states that there are 1,066,000 pupils enrolled in vocational courses approved by the Federal Board. The same report indicates that 20,779 teachers were employed in Federally aided schools during that year (1928). It is reasonable to expect that this number will increase as time goes on and the demands of an ever increasing population for satisfying its needs calls upon industry to provide maximum production at minimum cost. The result will be, as it always has been, a greater demand for adequately trained workers. The various agencies of industrial training must then be prepared to care for this actual and pressing need.

The Schools, a Telic Agency in a Democracy.—In a society, such as ours, based upon an industrial economy whereby the greater portion of the population engages in pursuits of an industrial, commercial, agricultural, or professional nature, it is reasonable to expect that the public schools, the great forces of social control, will assume the responsibility for training these people for their place in the social structure. The schools must do more than merely transmit the heritage of the past; they must serve as a telic agency in creating a firmer, finer, and happier life for the majority of the people. It is only with a knowledge of the roots from which the development of vocational training springs that a full and effective blossoming of that training can be obtained.

This blossoming will present an adequate program of vocational training directed toward economic efficiency and balanced by a recognition of the broader and wider interests of society. Chapman and Counts in their epitome, "Principles of Education," page 243, state:

"Such a program must seek (1) to increase production; (2) to secure an equitable distribution of goods and services to the masses of the people; (3) to foster wise and temperate consumption of these benefits; (4) to conserve those basic natural resources on which economic life depends; (5) to organize industry so that it will quicken rather than destroy the intellectual and moral life; and, finally, (6) to inject into industry a new spirit which will call forth the will to serve in place of the will to exploit."

CONCLUSION

Conclusion.—It is certainly evident that the influences of the various forces, which have been considered as roots of vocational education in this book tend toward an increasing realization of these six points.

It is the hope that some future study may be made that will employ the roots found in this book as a basis upon which future systems of vocational education may be organized. It is reasonable to expect that the roots which have contributed so much in the past will continue to effect future growth in much the same manner. Education, as a great social force, must lead the way in the creation and establishment of the agencies of vocational training. Educators and industrialists will be materially aided in their struggle to provide means for anticipated demands when they have a clear conception of the roots that have been instrumental in the development of present day practices.

Part Nine

CLASSIFIED BIBLIOGRAPHY

CLASSIFIED BIBLIOGRAPHY

I. History
- A. General Histories
    1. Ancient times
    2. Medieval times
    3. Modern times
- B. Economic and Industrial Histories
- C. Histories of Labor
- D. Histories of Education (general)
- E. Histories of Philosophy
- F. Philosophies of History

II. Economics
- A. General
- B. Money and Banking
- C. The Labor Problem
- D. Immigration
- E. Industrial Studies
- F. Miscellaneous

III. Social Studies

IV. Educational Studies
- A. Vocational
- B. Historical
- C. Philosophical

V. Legislative Studies and Acts

I. HISTORY

A. GENERAL HISTORIES

1. Ancient times:

Botsford, George, and Sihler, Ernest. Hellenic Civilization. Columbia University Press, 1915.

Osborn, Henry Fairfield. Men of the Old Stone Age. Scribner's, 1919.

Robinson, James, and Breasted, James. History of Europe.—Ancient. Ginn & Company, 1920.

Robinson, James Harvey. Readings in European History, Vol. I. Ginn & Company, 1904.

Van Hook, La Rue. Greek Life and Thought. Columbia University Press, 1924.

West, Willis M. Early Progress. Allyn & Bacon, 1920.

2. Medieval times:

Emerton, E. An Introduction to the Study of the Middle Ages. Ginn & Company, 1917.

Hackett, Francis. Henry the Eighth. Horace Liveright, 1929.

Henderson, Ernest. Select Historical Documents of the Middle Ages. Bell & Company, 1892.

Thorndyke, Lynn. Science and Thought in the Fifteenth Century. Columbia University Press, 1927.

3. Modern times:

Beard, Charles and Mary. The Rise of American Civilization. Macmillan, 1927.

Muzzey, David. The American People. Ginn & Company, 1927.

Osgood, H. L. The American Colonies in the Seventeenth Century. Macmillan, 1904.

Osgood, H. L. The American Colonies in the Eighteenth Century. Columbia University Press, 1924.

B. Economic and Industrial Histories

Bogart, E. L. The Economic History of the United States. Longmans Green, 1921.

Coman, Katherine. Industrial History of the United States. Macmillan, 1912.

Cowdrick, Edward. Industrial History of the United States. Ronald Press, 1923.

Droppers, Garrett. Economic History in the Nineteenth Century. Ronald Press, 1923.

Fisher, Irving. Resources and Industries of the United States. Ginn & Company, 1919.

Mac Gregor, David. The Evolution of Industry. Henry Holt, 1912.

Wells, Louis. Industrial History of the United States. Macmillan, 1922.

C. Histories of Labor

Beard, Mary. A Short History of the American Labor Movement. Harcourt Brace, 1920.

Brissenden, Paul. The I. W. W. Columbia University Press, 1920.

Carlton, F. T. The History and Problems of Organized Labor. D. C. Heath, 1911.

Carlton, F. T. Organized Labor in American History. D. Appleton, 1920.

Commons, John. History of Labor in the United States. Macmillan, 1918.
Ely, Robert T. The Labor Movement in America. Macmillan, 1905.
Gompers, Samuel. Seventy Years of Life and Labor. E. P. Dutton, 1925.
Hillquit, Morris. History of Socialism in the United States. Funk-Wagnalls, 1903.
Hoxie, R. F. Trade Unionism in the United States. Appleton, 1917.

D. Histories of Education (General)

Anderson, Lewis. History of Manual and Industrial School Education. D. Appleton, 1926.
Bennett, Charles A. History of Manual and Industrial Education up to 1870. The Manual Arts Press, 1926.
Hudson, J. W. The History of Adult Education. Longmans Green, 1851.
Monroe, Paul. A Textbook in the History of Education. Macmillan, 1907.
Monroe, Paul. Cyclopedia of Education. Macmillan, 1911.
Struck, F. Theodore. Foundations of Industrial Education. John Wiley Sons, 1930.

E. Histories of Philosophy

Durant, Will. The Story of Philosophy. Simon & Schuster, 1926.

F. Philosophies of History

Seligman, Edwin. The Economic Interpretation of History. Columbia University Press, 1924.
Spengler, Oswald. The Decline of the West. Knopf, 1926.

II. ECONOMICS

A. General

Carver, T. N. Principles of National Economy. Ginn & Company, 1921.
Ely, R. T. Outlines of Economics. Macmillan, 1917.
Smith, Adam. Wealth of Nations. Strahan, London, 1776.

B. Money and Banking

Goldenweiser, E. The Federal Reserve System in Operation. McGraw-Hill, 1925.
Harding, W. P. C. Formative Period in the Federal Reserve System. Houghton, 1925.
Hepburn, A. B. History of Currency in the United States. Macmillan, 1924.
White, H. Money and Banking. Ginn & Company, 1914.
Willis, H. P. The Federal Reserve System. Ronald Press, 1923.

C. The Labor Problem

Adams, T. S., and Sumner, H. L. Labor Problems. Macmillan, 1905.
Blum, Solomon. Labor Economics. Holt, 1925.
Commons, John R. Industrial Government. Macmillan, 1921.
Commons, John R., and Andrews. Principles of Labor Legislation. Harpers, 1916.
Lescohier, D. D. The Labor Market. Macmillan, 1918.
Lauck, William, and Sydenstricker, Edgar. The Conditions of Labor in American Industries. Funk-Wagnalls, 1917.
Marot, Helen. American Labor Unions. Henry Holt, 1914.
Marot, Helen. The Creative Impulse in Industry. E. P. Dutton, 1918.
Tannenbaum, F. The Labor Movement, Its Conservative Functions and Social Consequences. G. P. Putnam, 1921.
U. S. Commissioner of Labor. Annual Report, 1902. Annual Report, 1910.
Veblen, Thornstein. The Instinct of Workmanship. Macmillan, 1917.
Watkins, Gordon. An Introduction to the Study of Labor Problems.

D. Immigration

Fairchild, Henry P. Immigration. Macmillan, 1913.
Hourwich, I. A. Immigration and Labor. Putnam, 1912.
Jenks, J., and Lauck, W. J. The Immigration Problem. Funk-Wagnalls, 1913.

E. Industrial Studies

American Engineering Council. Report on Waste in Industry. McGraw-Hill, 1922.
Berglund, Abraham. The United Steel Corporation. Columbia University Press, 1907.
Bowden, W. Industrial Society in England Towards the End of the Eighteenth Century. Macmillan, 1925.
Foster, William Z. The Great Steel Strike. Viking Press, 1920.
Gulick, Charles. The Labor Policy of the United States Steel Corporation. Columbia University Press, 1924.
Interchurch World Movement. Public Opinion and the Steel Strike. Harcourt, 1921.
Knauth, Oswald. The Policy of the United States Towards Industrial Monopoly. Columbia University Press, 1913.

F. Miscellaneous

Cowdrick, Edward. Manpower in Industry. Holt, 1924.
Gregory, Keller, and Bishop. Physical and Commercial Geography. Ginn & Company, 1910.

Huntington, Ellsworth, and Cushing, Sumner. Principles of Human Geography. John Wiley, 1924.
Thurman and Van Meter. Principles of Railroad Transportation. Macmillan, 1918.
Usher, Abbot. A History of Mechanical Inventions. McGraw-Hill, 1929.

III. SOCIAL STUDIES

Abbott, Edith. Women in Industry. Appleton, 1910.
Cooley, Charles H. The Social Process. Scribner, 1926.
Ellwood, Charles. The Psychology of Human Society. Appleton, 1925.
Kelley, Florence. Modern Industry in Relation to the Family, Health Education, and Morality. Longmans Green, 1914.
Kelley, Florence. Some Ethical Gains Through Legislation. Macmillan, 1905.
Patterson, S. Howard. Social Aspects of Industry. McGraw-Hill, 1929.
Ross, E. A. Outlines of Sociology. Century, 1923.
Ross, E. A. Social Control. Macmillan, 1910.
Shenton, Herbert. The Practical Application of Sociology. Columbia University Press, 1929.
Smith, W. R. Introduction to Educational Sociology. Houghton Mifflin, 1925.

IV. EDUCATIONAL STUDIES

A. Vocational

Allen, Frederick. Principles and Problems in Vocational Guidance. McGraw-Hill, 1929.
American Federation of Labor. Education for All. Pamphlet, 1922.
Bradburn, William. Industrial Work in English Elementary Schools. Manual Training Magazine, 1918–19.
Carlton, F. T. Education and Industrial Evolution. Macmillan, 1908.
Carnegie Foundation for the Advancement of Teaching. Federal Aid to Vocational Education (by I. L. Kandall—Bulletin #10).
Chamber of Commerce of the United States. Referendum on the Report of the Committee on Education Regarding Federal Aid for Vocational Education. Pamphlet #14, 1913.
Commons, John R. Industrial Education and Dependency. University of Wisconsin Bulletin, General Series, #705, 1918.
Curoe, Philip. Educational Attitudes and Policies of Organized Labor in the United States. Teachers' College Contributions to Education, #201.
Davis, J. J., and Wright, J. C. You and Your Job. John Wiley, 1930.
Department of Labor, Canada. Vocational Education. Bulletin #28, August, 1928.

Douglas, P. H. American Apprenticeship and Industrial Education. Columbia University Press, 1921.

Eaton, Theodore. Education and Vocations. John Wiley, 1926.

Fleming, Arthur, and Pearce, J. G. The Principles of Apprenticeship Training. Longmans Green, 1916.

Gompers, Samuel. Industrial Education and the American Federation of Labor. Manual Training and Vocational Education, Peoria, Ill., Vol. 16, 1915.

Hart, Joseph. Adult Education. Crowell, 1927.

Holt, W. Stull. The Federal Board for Vocational Education. Appleton, 1922.

Keller, Franklin. Day Schools for Young Workers. Century, 1924.

Kelly, Roy. Training Industrial Workers. Ronald Press, 1920.

Kitson, H. D. The Psychology of Vocational Adjustment. Lippincott, 1925.

Leavitt, Frank. Examples of Industrial Education. Ginn and Company, 1912.

Leavitt, Frank. New Problems and Developments in Vocational Education. N. E. A. Proceedings (pp. 196–199), 1919.

Lee, Edwin. Objectives and Problems of Vocational Education. McGraw-Hill, 1929.

Link, Henry. Education and Industry. Macmillan, 1923.

Massachusetts Commissioner on Industrial and Technical Education, Report of, 1906. Columbia University Reprints, #1.

Morris, John V. Employee Training. McGraw-Hill, 1921.

Mulcaster, Richard. Positions. Longmans Green, 1888.

New York Industrial Education Survey. Board of Estimate and Apportionment, 1918.

Pabst, Alvin. Handwork Instruction for Boys. Manual Arts Press, 1910.

Prosser, Charles. Sociological Phases of Industrial Education. N. E. A. Proceedings, 1912.

Prosser and Allen. Vocational Education in a Democracy. Century, 1925.

Roman, Frederick. Industrial and Commercial Schools of the United States and Germany. Putnam, 1915.

Roman, Frederick. The New Education in Europe. Dutton, 1924.

Sears, William P. Training for Industry. New York University Thesis, 1925.

Smith, Homer. Industrial Education. Century, 1927.

Snedden, David. Administration and Educational Work of American Juvenile Reform Schools. Columbia University Press, 1907.

Snedden, David. The Problem of Vocational Education. Houghton Mifflin, 1910.

Snedden, David. Vocational Education. Macmillan, 1923.

CLASSIFIED BIBLIOGRAPHY

Society of Industrial Engineers, Industrial Education. S. I. E. Publications, Vol. 4, #1, 1921.
Stormzand, Martin J. Progressive Methods of Teaching. Houghton Mifflin, 1927.
Strong, Edward, and Uhrbrock, Richard. Job Analysis and the Curriculum. Williams and Wilkins, 1923.
Swift, Fletcher. Federal Aid to Public Schools. U. S. Bureau of Education, 1922.
Vaughan, S. J., and Mays, Arthur. Content and Method in Industrial Arts. Century, 1924.
Wright and Allen. Administration of Vocational Education. John Wiley, 1929.
Wright and Allen. Efficiency in Education. John Wiley, 1929.
Wright and Allen. Supervision of Vocational Education. John Wiley, 1926.

B. Historical

Bache, Alexander. Report on Education in Europe to the Trustees of Girard College for Orphans. Philadelphia, 1839.
Baker, Ray Palmer. A Chapter in American Education. Scribner, 1924.
Balfour, Graham. The Educational Systems of Great Britain and Ireland. Oxford, 1903.
Barnard, Henry. American Journal of Education, 1858, 1862.
Barnard, Henry. National Education in Europe. Norton, 1854.
Bartley, George. The Schools of the People. Bell and Daldy, 1871.
Bennett, Charles. A History of Manual and Industrial Education up to 1870. Manual Arts Press, 1926.
Brougham, Lord Henry. A System of Education on the Principle of Connecting Science with Useful Labor. Edinburgh Review #61, 1918.
Burns, C. Delisle. A Short History of Birkbeck College. University of London Press, 1924.
Coates, Charles. History of the Manual Training Schools of Washington University. U. S. Bureau of Education, Bulletin #3, 1923.
Davies, J. L. The Workingman's College. Macmillan, 1904.
Griscom, John. A Year in Europe (1818–19). Collins, 1923.
Guthrie, Thos. Seed Time and the Harvest of the Ragged Schools. Black, 1860.
Hodge, George. Association (Y. M. C. A.) Education Work. Association Press, 1912.
Hubbell, George. Horace Mann in Ohio. Columbia University Press, 1900.
Inglis, A. The Rise of the High School in Massachusetts. Teachers' College Dissertation #45, 1911.
Jackson, G. L. Developments of School Support in Colonial Massachusetts. Teachers' College Dissertation #25, 1909.

Kay, Joseph. The Social Conditions and Education of the People of England and Europe. Longmans Green, 1850.
Kemp, W. W. Support of Colonial New York Schools by Society. Teachers' College Dissertation #56, 1913.
Knight, F. B. Tendencies in Secondary Education in England and the United States. Teachers' College Dissertation #120, 1922.
Lockwood, George. The New Harmony Movement. Appleton, 1905.
McGrath, John. A History of Vocational Education. New York University Dissertation, 1908.
Mays, Arthur. The Determining Factors in the Evolution of the Industrial Arts in America. Bruce Publishing Company.
Mead, A. R. The Development of Free Schools in the United States. Teachers' College Dissertation #91, 1918.
Monroe, Paul. A Textbook in the History of Education. Macmillan, 1907.
Monroe, Paul. Cyclopedia of Education. Macmillan, 1911.
Montague, E. F. Sixty Years in Waifdom, or the Ragged School Movement in English History. Murray, 1904.
Owen, Robert Dale. Threading My Way. Carleton, 1874.
Playfair, Lyon. Industrial Instruction on the Continent. Longmans, 1852.
Prosser, Charles. A Study of the Boston Mechanic Arts High School. Teachers College Dissertation # 74, 1915.
Ricketts, Palmer. A History of Rensselaer Polytechnic Institute. John Wiley, 1914.
Sadler, M. E. Continuation Schools in England and Elsewhere. University Press, Manchester, 1908.
Scott, Jonathan. Historical Essays on Apprenticeship and Vocational Education. Ann Arbor Press, 1914.
Seybolt, Robert. Apprenticeship and Apprenticeship Training in Colonial New England and New York. Teachers' College Dissertation #85, 1917.
Seybolt, Robert. The Evening School in Colonial America. University of Illinois, 1925.
Snedden, David. The Birth and Childhood of Vocational Education. N. E. A. Address, 1918.
Westermann, W. L. Vocational Training in Antiquity. School Review, University of Chicago, 1914.
Woodbridge, William. Sketches of Hofwyl. American Annals of Education, Boston, 1831.
Woody, Thomas. Early Quaker Education in Pennsylvania. Teachers College Dissertation #165, 1920.

C. Philosophical

Basedow, Johann. Nouvelle Méthode d'Education. Huber, 1772.
Bode, Boyd. Fundamentals of Education. Macmillan, 1921.
Bode, Boyd. Modern Educational Theories. Macmillan, 1927.

Dewey, John. Democracy and Education. Macmillan, 1916.
Dewey, John, and Dewey, Evelyn. Schools of Tomorrow. Dutton, 1915.
Francke, A. H. Faith's Work Perfected. Gage Translation, London, 1867.
Froebel, Frederick. The Education of Man. Appleton, 1905.
Herbart, Johann. The Science of Education. Sonnenschein Company, 1892.
Martin, Everett D. The Meaning of a Liberal Education. W. W. Norton, 1926.
Pestalozzi. How Gertrude Teaches her Children. Bardeen Edition, 1898.
Rousseau, Jean Jacques. Emile.

V. LEGISLATIVE STUDIES AND ACTS

Ayusawa, I. International Labor Legislation. Columbia University Press, 1920.
Baker, Elizabeth. Protective Labor Legislation. Columbia University Press, 1925.
Federal Statutes. Hatch Act, 1887, 1906. Smith-Lever Act, 1914. Morrill Act, 1862. Smith-Hughes Act, 1917. Smith-Bankhead Act, 1920. Sheppard-Towner Act, 1921.
French Commission on Technical Training. Eyre & Spottiswoode, 1868.
Groat, George C. Attitude of American Courts in Labor Cases. Columbia University Press, 1911.
Ogburn, William. Progress and Uniformity in Child Labor Legislation. Columbia University Press, 1912.

Books listed in the bibliography may be obtained from the publishers indicated. The addresses are as follows:

Allyn and Bacon, Boston, Mass.
D. Appleton and Company, New York, N. Y.
Association Press (Y. M. C. A.), New York, N. Y.
G. Bell and Sons, Ltd., London, England.
Adam and Charles Black, Edinburgh, Scotland.
The Bruce Publishing Company, Milwaukee, Wis.
C. W. Carleton and Company, New York, N. Y.
The Century Company, New York, N. Y.
Collins and Company, New York, N. Y.
Columbia University Press, New York, N. Y.
Thomas Y. Crowell Company, New York, N. Y.
E. P. Dutton, New York, N. Y.
Eyre and Spottiswoode, London, England.
Funk and Wagnalls Company, New York, N. Y.
Ginn and Company, New York, N. Y.
Government Printing Office, Washington, D. C.
Harcourt, Brace and Company, New York, N. Y.

Harper and Brothers, New York, N. Y.
D. C. Heath, Boston, Mass.
Henry Holt and Company, New York, N. Y.
Houghton Mifflin Company, Boston, Mass.
Alfred A. Knopf, New York, N. Y.
J. B. Lippincott Company, Philadelphia, Penna.
Horace Liveright, New York, N. Y.
Longmans Green & Company, New York, N. Y.
The Macmillan Company, New York, N. Y.
The Manual Arts Press, Peoria, Ill.
McGraw-Hill Book Company, New York, N. Y.
Charles Murray Company, London, England.
Charles Norton, New York, N. Y.
Oxford University Press, New York, N. Y. (American Branch).
G. P. Putnam's Sons, New York, N. Y.
The Ronald Press Company, New York, N. Y.
Charles Scribner's Sons, New York, N. Y.
Simon and Schuster, New York, N. Y.
Sonnenschein Company, London, England.
Teachers College Contributions to Education, Columbia University, New York, N. Y.
University of California Press, Berkeley, Calif.
University of Chicago Press, Chicago, Ill.
University of Illinois Press, Urbana, Ill.
University of London Press, London, England.
University of Manchester Press, Manchester, England.
Viking Press, Inc., New York, N. Y.
John Wiley and Sons, New York, N. Y.
Williams and Wilkins, Baltimore, Md.

INDEX

A

Address before the Council of Arts and Manufactures of the Province of Quebec, 230
Adult education, 207
Agricultural education, 155, 156
Alcott, Amos Bronson, 100
Alcuin, of York, 76, 79
Alfred the Great, 65, 79, 80
Alien Contract Immigration Law, 38
Allen, Charles R., 108, 200, 289
American Federation of Labor, 49, 123, 165, 166, 167, 168, 169, 170, 171, 198, 286
American Revolution, 20, 21
American Society of Mechanical Engineers, 205
Anderson, Lewis F., 161, 162, 164
Apprenticeship system, 67, 68, 80, 82, 106, 140, 142, 143, 162, 218, 219, 220, 221, 222, 280
Aristotle, 65, 72, 73
Arkwright, 10
Athenian concept of labor, 73
Athenian education, 70, 71, 72, 73, 74
Ayres, Leonard P., 185

B

Bache, Alexander, 129, 152
Bacon, Francis, 65, 84
Bennett, Charles A., viii, 63, 77, 78, 94, 95, 97, 99, 104, 129, 133, 134, 135, 139, 147, 148, 149, 152, 153
Bibliography, 293

Big business, 51, 211, 213, 214, 215, 216, 217, 218, 219, 220, 221, 222, 223, 224, 225
Binet and Simon, 183, 185
Birkbeck, George, 147, 148
Black Death in England, 6, 8
Bogart, E. L., 1
Boston Elevated Railway School, 214, 215, 216
Burgh schools, 65, 82
Business and vocational education, 211, 213, 214, 215, 216, 217, 218, 219, 220, 221, 222, 223, 224, 225

C

Canada, vocational education in, 227-240
Canal-building era, 28
Carling, the Honorable Mr., 229
Cartwright, 10
Cattell, J. McK., 182
Census of 1790, 23
Charlemagne, 65, 76, 79
Cheyney, E. P., 8
Chivalry and chivalric training, 80
Christian (early) education, 76
Civil War, 33, 37, 38, 39
Cokesbury College, 89
Collective bargaining, 12
Colonial pursuits, 16, 17, 18
Combination Acts, 12
Comènius, 84
Compromise of 1850, 36
Continental Congress, 22
Continuation schools, 114, 115, 281

INDEX

Cooperative system of education, 118, 119, 120, 121, 122, 123
Corporation school, 111, 112, 283
Correspondence schools, 112, 113
Council of National Defense, 55
Counts, George S., 183, 188, 189, 290
Cowdrick, Edward S., viii, 1, 17, 20, 25, 60
Crompton, Samuel, 10

D

Dalton Plan, 191, 192
Darwin, Charles, 181, 182
Democracy, 287, 290
Depression of 1922, 57, 58
Depression of 1929, 59
Development of the West, 27, 39
Dingley Tariff, 52
Division of labor, 4, 12, 66, 280
Domestic system, 7, 9

E

Ebbinghaus, Hermann, 182
Educational Act of 1918 (Great Britain), 142
Educators and vocational education, 180, 181, 182, 183, 184, 185, 186, 187, 188, 189, 208, 209, 210
Egyptian education, 68
Elimination in the schools, 186
Ely, Richard, 7
Emergency Fleet Corporation, 55
Enclosures, 6, 12
Erasmus, 65, 84
Erie Canal, 28
Evening schools, 113, 114, 166, 231
Experimental movement in education, 184

F

Factory Act of 1802, 140
Factory Act of 1819, 141
Factory Act of 1833, 141
Factory Act of 1844, 142
Factory schools, 111, 112, 211, 214, 215, 216, 217, 218, 219
Factory system, 9, 10, 13, 24, 34, 107, 279
Federal aid to vocational education, 124, 125, 126, 127, 128, 198, 199, 200, 201, 202, 289
Federal Board for Vocational Education, 105, 115, 127, 199, 200, 201, 202, 289
Federal Commission on National Aid to Vocational Education, 199
Fellenberg, Philippe Emanuel von, 65, 90, 91, 92, 93, 94, 101, 128, 136
Fess-Kenyon Act, 127, 128
Food Control Act, 56
Fordney-McCumber Tariff, 53
Francke, 84
Fraternal Associations and Vocational Education, 177, 286
French and Indian War, 21
Froebel, F. W. A., 94, 96
Fulton, Robert, 28
Future trends in vocational education, 241, 242, 243, 244, 245, 246, 247, 248

G

Galton, Francis, 182
General Electric Company Schools, 216, 217
George-Reid Act, 128
Girard College, 129, 138, 152
Gold discovered in California, 36
Golden age of English labor, 8
Greek education, 69, 70, 71, 72, 73, 74
Griscom, John, 91, 93, 132
Guilds, 7, 82
Guild schools, 82

INDEX

H

Hall, G. Stanley, 182
Hampton Institute, 139
Handicraft era, 7
Hargreaves, James, 10
Hatch Act, 126, 198
Hawley-Smoot Tariff, 54
Herbart, Johann Friedrich, 65, 95
Higher technical education, 119, 144, 151, 152, 153, 154, 155, 156, 157, 222, 245
Homestead Act, 40
Hours of labor, 14, 47, 48, 281
Howe, Elias, 31
Human engineering, 107, 213, 214, 283
Hundred Years' War, 6

I

Immigration, 33, 60, 61
Indentured servants, 18
Industrial arts, 116, 117
Industrial cooperation, 48, 50
Industrial evolution in England, 3
Industrial evolution in the United States, 16, 17, 18, 23, 24, 25, 30, 31, 32, 162, 213
Industrial expansion, 10, 30, 31, 32, 38, 44, 45, 46, 54, 57, 59, 107, 213
Industrial Revolution in England, 9, 10, 13, 98, 278
Industrial Revolution in the United States, 16, 17, 18, 23, 24, 25, 278
Industrial schools in early America, 88, 89, 138
Industrial schools for poor or delinquent children, 128, 129, 130, 131, 136, 285
Industrial system, 9, 10, 13, 16, 17, 18, 283
Industrial trends, v, 243, 244, 245, 278, 279, 280, 281, 284

Industry in early America, 16, 17, 18, 23, 24
I.W.W., 50
Inventions, 9, 31

J

James, William 182, 183
Jewish education, 68, 69
Job analysis, 192, 193, 194, 195
Job sheets, 192, 193, 194, 195

K

Kay, 10
Kansas Industrial Court, 48
Knights of Labor, 49

L

Labor in the 14th century, 8
Labor, types after 1870, 107
Labor and vocational education, 107, 123, 161, 165, 166, 167, 168, 169, 170, 171, 172, 286
Labor in the American Colonies, 18
Labor in the American Republic, 25, 26, 34, 38, 46, 107, 279
Labor legislation, 6, 8, 12, 13, 32, 48
Labor problem, 11, 12, 14, 26, 32, 34, 46, 50, 57
Labor unionism, 12, 26, 32, 38, 47, 48, 49, 50, 57
Laggards in our schools, 185
Laissez-faire, 10, 13, 14, 15
Leaders in the field of vocational education, 208, 209, 210
Legislation, a root of vocational education, 288
Localization of production centers, 244
Luther, Martin, 83
Lyceums, 151

M

Macy's employee training program (R. H. Macy Co.), 220, 221
Magazines devoted to vocational education, 205, 206
Manchester Board of Health Report, 14
Manorial system in England, 3, 4, 5, 6
Manual labor movement, 101, 102, 103, 104
Manual training, 108, 117, 118
Masonic endeavors in education, 177, 178
Massachusetts Commission on Industrial and Technical Education, 109, 161, 162, 163
Mays, Arthur B., 63, 106, 109, 112, 113, 117, 124, 125, 127
McCormick, Cyrus, 31, 288
McKinley Tariff, 52
Mechanics' institutes, 147, 148, 149, 150, 151
Mercantile system, 9, 20
Milestones in the development of vocational education, 249, 251
Milton, John, 84
Missouri Compromise, 35
Monastic life, 77, 78, 80
Monetary history, 20, 21, 22
Monroe, Paul, 66, 86
Montaigne, 65
Mooseheart, 179, 180
Morrill Act (Land Grant Colleges), 126, 156, 198, 288

N

National City Bank's Educational Trust Fund, 224, 225
National Educational Association, 163
National Grange and agricultural education, 221
National Labor Union, 49
National Metal Trades Association (employee training program), 217
National Society for the Promotion of Industrial Education, 110, 164, 199
Navigation Acts, 21
New York Central Railroad Company (employee training programs), 218, 219
Niemeyer, August, 65, 95
Norman conquest, 3

O

Oberlin, John, 97
Oneida Institute, 102, 103
Oregon, 36
Owen, Robert, 15, 65, 91, 98, 99

P

Panic of 1873, 39
Payne-Aldrich Tariff, 53
Peasants' revolt, 6, 8
Pestalozzi, Heinrich, 65, 85, 86, 87, 99, 100, 128, 136
Plato, 65, 72
Population of colonial America, 20
Pound, Arthur, 108
Primitive education, 65, 66, 67
Private trade school, 108, 110
Prosser, Charles A., 108, 200, 289

R

Rabelais, 65, 84
Radio Corporation of America (training program), 218
Ragged Schools of Great Britain, 133, 134, 136
Railroad development, 29, 39, 41, 42, 43
Railroad legislation, 43, 44

INDEX 309

Reconstruction era, 41
Reformation, 65, 83
Reform schools in America, 139
Renaissance, 65, 83
Rensselaer Polytechnic Institute, 153, 154, 155
Rogers, Ezekiel, 17
Roman education, 74, 75, 76
Roots of Vocational Education, a summary, 275
Rousseau, Jean Jacques, 65, 85, 207

S

Saracens, 81, 82
Schneider, Herman, 119
Scientific Progress, 181, 182, 183, 213, 284
Selective character of education, 188, 189
Selvidge, Robert W., 192
Sense-realism in education, 84
Sheppard-Towner Act, 127
Slater, Samuel, 24
Slavery, 19, 34
Sloyd, 118
Smith, Adam, 13, 94
Smith, Walter, 230
Smith-Hughes Act, 105, 123, 127, 128, 170, 200, 201
Smith-Lever Act, 127, 198
Snedden, David, 139, 203
Social agencies and vocational education, 161, 204
Society for the Promotion of Engineering Education, 205
Socrates, 65, 72
Sophists, 71
Spartan education 70, 71
Specialization in industry, 108
Stamp Act, 21
Statute of Apprentices, 8, 13
Statute of Laborers, 6, 8
Stephenson, George, 10

Strikes, 50
Sugar Act, 21
Sunday school, 145, 146

T

Tarbell, Ida, 53
Tariff History, 29, 51, 52, 53
Taussig, 52
Taylor, Fred, 213
Technical education, 119, 144, 151, 152, 153, 154, 155, 156, 157, 222, 223, 224, 245
Terman, L., 183, 184
Testing movement in education, 183, 184
Texas Annexation, 36
Textile industries (United States), 24
Thorndike, Edward L., 3, 182
Trade schools, 108, 109, 124, 173
Trade tests, 184
Transportation, 27, 28, 29
Tully, Christopher, 24

U

Underwood Tariff, 53
United States Shipping Board, 55
Universities in the Middle Ages, 81
University of California, Publications on Vocational Education, 195, 196

V

Van Denburg, Joseph, 185, 186, 187, 188
Vocational education defined, vi
Vocational Education Act (see Smith-Hughes Act)
Vocational education and the future, 241, 243, 244, 245, 246, 247, 248

INDEX

Vocational education and the public schools, 110
Vocational education for girls, 92, 105, 123, 124, 223
Vocational education for Negroes, 138
Vocational education in Canada, 227
Vocational schools (classified), 105
Voltaire, 85

W

Wages, 282
War Industries Board, 56
War Labor Board, 57
War of 1812, 23
Watkins, Gordon, 11, 15
Watt, James, 10
Wehrli, 91, 93, 96, 97, 136

Western (United States) development, 27, 39
Westinghouse Electric and Manufacturing Company—Training Programs, 221, 222, 223, 224
Whitney, Eli, 10, 24
Winnetka Plan, 191, 192
Women in industry, 282
Woodworth, Robert S., 181, 182, 183
World War, 54, 55, 56, 170
Wright, J. C., ix
Wundt, Wilhelm Max, 181

Y

Yale and Towne Manufacturing Company School, 219, 220
Young Men's Christian Association, 172, 173, 174, 175

NOV 23 1990